ISBN 978-0-265-93147-9
PIBN 10911672

This book is a reproduction of an important historical work. Forgotten Books uses
state-of-the-art technology to digitally reconstruct the work, preserving the original format
whilst repairing imperfections present in the aged copy. In rare cases, an imperfection in
the original, such as a blemish or missing page, may be replicated in our edition. We do,
however, repair the vast majority of imperfections successfully; any imperfections that
remain are intentionally left to preserve the state of such historical works.

For support please visit www.forgottenbooks.com

DEPARTMENT OF COMMERCE

TECHNOLOGIC PAPERS

OF THE

BUREAU OF STANDARDS

S. W. STRATTON, Director

No. 121

STRENGTH AND OTHER PROPERTIES
OF WIRE ROPE

BY

J. H. GRIFFITH, Associate Engineer Physicist

and

J. G. BRAGG, Assistant Physicist

Bureau of Standards

ISSUED JULY 16, 1919

PRICE, 20 CENTS

Sold only by the Superintendent of Documents, Government Printing Office,
Washington, D. C.

WASHINGTON
GOVERNMENT PRINTING OFFICE
1919

STRENGTH AND OTHER PROPERTIES OF WIRE ROPE

By J. H. Griffith and J. G. Bragg

CONTENTS

I. INTRODUCTION

1. PURPOSE OF TESTS

There have been few systematic researches conducted by engineering laboratories to determine the physical properties of wire ropes. The tests which have been made by manufacturers are, as a rule, not available for critical comparative study by engineers. The investigations which have been made abroad, notably those by Tetmajer and the South African Commission, have covered particular types of constructions, such as cables for tramway and mine hoists. The results can not be strictly applied to American practice. The reason that systematic experimentation in this field has been somewhat limited may be attributed to the fact that it is difficult to obtain a large number of specimens for test purposes which have been selected under uniform specifications. The relative cost of preparing specimens is, moreover, as a rule, quite out of proportion to the yield of test data. A considerable range of variation may be expected in the observed data on different specimens, so that a larger number of test specimens is requisite in obtaining appropriate averages of physical properties than is ordinarily required in other tests upon the materials of construction.

It is the purpose in this paper to give a digest of the results of tests of about 300 cables selected under the specifications of the Isthmian Canal Commission. The specimens were submitted primarily for the purpose of fulfilling acceptance tests upon material used at the Canal Zone. The tensile strength of the specimens was the important consideration, but the major portion of the investigation has been of a purely supplementary character to determine the laws of behavior of the cables in connection with their important physical characteristics.

2. MANUFACTURERS REPRESENTED

The cables to be described were submitted from the plants of the following manufacturers: The Broderick & Bascom Rope Co., St. Louis, Mo.; A. Leschen & Sons Rope Co., St. Louis, Mo.; Macomber & Whyte Rope Co., Chicago, Ill.; Hazard Manufacturing Co., Wilkes-Barre, Pa.; Wright Wire Co., Palmer, Mass.; Waterbury Co., New York, N. Y.; John A. Roebling's Sons Co., Trenton, N. J.; and American Steel & Wire Co., several plants.

It seemed important to treat the manufacturer as a variable of the investigation. It was felt, however, that it would be unjust to draw any conclusions from the comparative test data in this respect without giving at the same time the fullest description of processes of manufacture and particular grades of steel used, trade names, etc. It was considered that any needs of the investigation in accounting for a possible uniformity of results with respect to one manufacturer's product would be served by indicating the manufacturer impersonally by an appropriate symbol. In the tables the manufacturer is designated by a letter with a suitable numeral as M–9, etc., without reference to the list above given. No other identification is given, and trade names are omitted. The particular grade or quality of any one type of steel or other material is to be inferred from the test data.

3. PERSONNEL OF INVESTIGATION

The investigation was started in 1908 at the structural materials laboratory of the Geological Survey. Acknowledgments are due to N. D. Betts, W. C. Campbell, H. Kaplan, L. H. Losse, E. R. Gates, and T. N. Holmes for some of the earlier work which was performed under the direction of Richard L. Humphrey. The laboratory was placed under the administration of the Bureau of Standards in 1910. The authors have continued the tests up to the present time, and are responsible for the work of collation of the data.

II. CONSTRUCTION AND CLASSIFICATION OF TEST SPECIMENS

1. GENERAL CONSTRUCTION OF CABLES DESCRIBED

It has been found in the development of the wire-rope industry that certain arrangements of wires in a cable strand afford more stable combinations and are otherwise more efficient in meeting the provisions of specifications than others. Manufacturers, as a the result of their experience, have adopted standard types of construction [1] and have used particular grades of steel to best fulfill the needs of engineering practice. One type of cable, for example, is more applicable where static strength is the important factor and another where a high abrasive resistance is to be developed. One type is better fitted for power-transmission purposes and another for ship riggings, as the case may be.

A cable is composed of strands. The strand is the fundamental unit of its construction. The wires of these strands are twisted together symmetrically according to a definite geometrical arrange-

---

[1] Reference may be made to the handbooks and trade catalogues issued by the manufacturers.

ment.   One wire is placed at the center of the strand in ordinary
construction.   This wire is surrounded with successive concentric
rings of wires containing 6,12,18, and 24 or more wires according to
the type used.   (See Fig. 1.)   The cables in this investigation
have either 6 or 8 strands with different arrangements of wires,

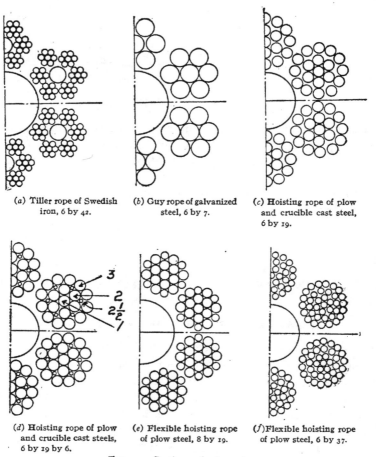

(a) Tiller rope of Swedish
iron, 6 by 42.

(b) Guy rope of galvanized
steel, 6 by 7.

(c) Hoisting rope of plow
and crucible cast steel,
6 by 19.

(d) Hoisting rope of plow
and crucible cast steels,
6 by 19 by 6.

(e) Flexible hoisting rope
of plow steel, 8 by 19.

(f)Flexible hoisting rope
of plow steel, 6 by 37.

FIG. 1.—*Sections of wire rope*

First numeral in 6 by 19 by 6 of (d) refers to number of strands, second to number of wires in a strand,
and third to number of filler wires in a strand. Other numerals of (d) refer to location of "rings" of wires
and similarly for the other sections.

which will be described later in detail.   The construction of the
cable is briefly specified by giving the number of strands in the
cable and the number of wires in a strand.   For example, a cable
having 6 strands of 19 wires each, as in (c) of the figure, is briefly
described as a 6 by 19 construction.   Sometimes additional filler

wires are inserted in such a way as to reduce the open spaces between the wires. A third figure is then added and the construction is indicated, for example, as 6 by 19 by 6, as in (*d*) of the figure.

The strands are grouped about a rope core of manila or other suitable fiber, which is effective in holding a lubricant for the wires and also in providing an appropriate bedding for the strands. Empirical equations expressing the general laws of the construction for the different types of cables will be given later in this report.

2. CLASSIFICATION AND SPECIFICATIONS

The classification of the test specimens given in this paper is purely an arbitrary one. Manufacturers make numerous other types of cables than those to be discussed. The classification to be described was selected because it follows the main subdivisions given in the specifications. It was found to be useful in the arrangement and grouping of the test data for analysis and discussion. While certain particular types are not included in the report, it is believed that in any large engineering construction operation the relative number of cables used of each type and diameter and the weight of importance which attach to those types will bear some approximate relation to those given in the following classification. A few results of tests of larger-size cables conducted at the Washington laboratory, and of other cables not given in the classification, have been given in the report as matters of general interest.

(*a*) *Tiller Rope.*—This is the most flexible type of cable manufactured. Such cables are used where the loading is light and bending over small sheaves is required, as in the case of boat tillers. They are not adapted to resisting surface abrasion on account of the small diameter of the wires.

The cables tested of this type are of Swedish iron. The construction is 6 by 6 by 7. (See Fig. 1 (*a*).) The diameters range from ¼ inch to 1 inch. It was stipulated in the specifications that the material was to be used on small boats and for similar services where extreme flexibility is necessary. "Rope is to be made from high-grade Swedes iron stock. Rope is to be composed of 252 wires, made up of a hemp core, around which are twisted 6 ropes, each of which consists of 6 strands inclosing a hemp center; each strand to have 7 wires."

The tensile strength to be developed for tiller rope was **not** mentioned. The strengths specified by one of the manufacturers for iron tiller rope are as follows:

Diameter	Tensile strength	Diameter	Tensile strength
Inches	Pounds	Inches	Pounds
1/4	1300	5/8	7000
3/8	3000	3/4	11 000
1/2	5800	1·	22 000

(b) *Guy Rope.*—This rope is used for the guying of steel stacks, derrick masts, and gin poles in engineering construction work, for ship rigging, etc., where there is static loading without bending on sheaves and little impact. The wires are usually galvanized to resist weathering and corrosive vapors. The construction is 6 by 7. (See Fig. 1 (b).) Since there are comparatively few wires and these are relatively of large diameter, the 6 by 7 construction is the least flexible type of rope. It is sometimes used for haulage purposes, where the cables are **not** bent over sheaves. It is well fitted, on account of the relative size of the wires, to resist surface abrasion.

The specifications called for "galvanized iron or steel standing rope to be used in connection with ship rigging, guys for derricks, guys for smokestacks, etc. Rope is to be coarse laid and composed of 6 strands of 7 wires to the strand. * * *. Wire shall be well galvanized and shall be what is known to the trade as extra galvanized."

It was stated in the specifications that these ropes shall have a minimum tensile strength, as follows:

Diameter	Tensile strength	Diameter	Tensile strength
Inches	Pounds	Inches	Pounds
3/8	3900	1	28 200
1/2	6800	1 1/8	36 000
5/8	11 400	1 1/4	46 000
3/4	15 600	1 3/8	52 000
7/8	22 200	1 1/2	60 000

The values in the above table range from 5 to 12 per cent below the standard strengths adopted by a committee of the manufacturers in May 1910 for iron rope of this class. The ropes tested were galvanized steel.

(c) *Hoisting Rope of Crucible Cast Steel.*—This rope was not mentioned in· the specifications, but was submitted for testing. It is commonly used for mine hoists, elevators, conveyors, derricks, and kindred purposes. Crucible cast steel rope possesses about double the strength of iron rope of the same diameter. Crucible steel is described by the manufacturers as an acid open-hearth carbon steel. In the finished wire it has a tensile strength varying from 150 000 to 200 000 pounds per square inch. The ropes tested are of the 6 by 19 and 6 by 19 by 6 construction, as shown in Fig. 1 (c) and (d).

The 1910 standard strengths adopted by the committee of manufacturers for this class are as follows:

Diameter	Tensile strength	Diameter	Tensile strength
Inches	Pounds	Inches	Pounds
1/4	4400	7/8	46 000
3/8	9600	1	60 000
1/2	16 800	1 1/8	76 000
5/8	25 000	1 1/4	94 000
3/4	35 000	1 3/8	112 000
		1 1/2	128 000

(d) *Hoisting Rope of Plow Steel.*—The specifications stated that the rope was to be used on locomotive and wrecking cranes and for similar heavy work. The ropes tested are of the 6 by 19 and 6 by 19 by 6 construction, as shown in Fig. 1 (c) and (d).

Plow steel is described by the manufacturer as an acid open-hearth medium-high carbon steel, having a tensile strength in the finished wire varying from 220 000 to 260 000 pounds per square inch, this depending somewhat on the size of the wires and the particular grade of plow steel. It was stated in the specifications that this rope should possess a minimum tensile strength for different diameters, as follows:

Diameter	Tensile strength	Diameter	Tensile strength
Inches	Pounds	Inches	Pounds
3/8	11 500	1	76 000
1/2	20 000	1 1/8	94 000
5/8	31 000	1 1/4	116 000
3/4	46 000	1 3/8	144 000
7/8	58 000	1 1/2	164 000

It was specified that the wire used in the construction should have an elongation in 8 inches of about 2½ per cent. The above tensile strengths coincide with the standard strengths adopted by the manufactures May, 1910.

(e) *Extra-Flexible Hoisting Rope of Plow Steel.*—It was specified that this class of rope was to be used "in connection with steam-shovel swinging gear and similar service, where it is wound on small diameter drums." The rope is of 8 by 19 construction, indicated in Fig. 1 (e). It was stated that 6 by 37 construction might be substituted for rope having a larger diameter than 1 inch. The minimum tensile strength to be developed was given in the specifications as follows:

Diameter	Tensile strength	Diameter	Tensile strength
Inches	Pounds	Inches	Pounds
1/4	4500	1	66 000
3/8	10 240	1 1/8	86 000
1/2	17 400	1 1/4	104 000
5/8	28 000	1 3/8	128 000
3/4	40 000	1 1/2	148 000
7/8	52 000		

These tensile strengths are the same as the standard strengths for plow-steel cables of this class adopted by the manufacturers in 1910.

The specifications also stated that the wire entering into the construction should develop an elongation in 8 inches of about 2½ per cent.

III. SCOPE OF THE INVESTIGATION

It is the intention in this paper to discuss the physical charac-teristics of the cables as submitted from the results of the tests. The laws of arrangement of the strand and wires and the relations which exist between the diameter of the cables, their constituent wires, rope cores, and the pitches of the wires and strands have been determined.

Analyses are submitted of the steel, hemp fibers, and lubricants of plow-steel cables. These show the grades of material used and the variations that may exist for cables of the same class submitted by different manufacturers. The variations found are doubtless typical of those which exist for the other classes.

The maximum loads and stresses developed by tensile tests have been found for all the specimens. The types of fractures have been recorded in each case to show, if possible, a relation between the maximum load and the manner of failure of the specimen.

Stress-strain measurements were made upon over 50 per cent of the cables tested to determine the percentage of elongation and the lateral contraction of the specimens under cumulative loads. These data are important in developing a rational mechanics of the cable, and show to what extent a cable possesses elastic structure. The data have been used for determining the moduli of the cables. The modulus is employed in the calculation of flexural stresses when a cable is bent over a sheave for the transmission of power. These data may also be employed in investigating the bending moment and torque developed in a strand when it is analyzed as a helical spring.

The results of a large number of individual tests of wires have been presented to show the uniformity in the properties of steel employed for cables subject to kinetic loading. The wires for this purpose were taken from the specimens before the tensile tests were made. The mean tensile strengths and percentages of elongations in the wires were determined, and the amounts of variations are recorded for comparison with the elongations found for the cables.

A general analysis is given of the distributions of stress in the constituent wires of a cable. This has been employed for interpreting the modes of fracture of cables and the effects upon the strength of wide variations in the elongations of wires. The ratio of the strengths of cables to the strengths of their aggregate wires have been determined.

The results of the tests have been analyzed by statistical methods, and the conclusions as to the fundamental properties and laws of wire rope are stated.

IV. DETAILS OF CONSTRUCTION AND MEASUREMENTS OF CABLES

1. PRIMARY DATA FROM MEASUREMENTS

Tables 3 to 13, inclusive, give a list of 275 specimens upon which tests were conducted. Each cable is given a serial number, these being taken in numerical order. The classification, diameter, and other fundamental data are recorded in the tables under the heading "General data." The diameters recorded are the rated diameters of the manufacturers, and represent the diameters of the cylindrical envelope of the specimen instead of the lesser "diameter" of the prismoidal envelope inclosing strands. The specimens are arranged in the tables in the order of the groups as previously described in the classification; also in the order of increasing diameters.

(a) *Cross Sectional Areas of the Cables.*—In determining the cross-sectional areas the observer obtained the mean of several measurements of each diameter of the component wires of a single strand, using a Brown & Sharpe screw micrometer for the purpose, and from these diameters calculated the area of the wires, the sum of which when multiplied by the number of strands gives the aggregate area of the wires in the cable. All wires are included, including the filler wires which are sometimes used in the cable construction as in Fig. 1 (d). These cross-sectional areas were determined for each cable, and are given under the appropriate column of tables, together with the mean area for any particular group of cables found by averaging the results. The areas determined in this manner are the nominal areas commonly used in obtaining the approximate stress upon the cross sections.

(b) *Formulas for the Diameters of Wires and Sectional Areas of Cables.*—The mean sectional areas of the cables and the mean diameters of the wires for each group are given for ready reference in Table 1. The mean diameters of the wires were calculated from the mean areas by the formula

$$d = \left(\frac{A}{0.7854 \times n_c} \right)^{\frac{1}{2}}$$

where d is the mean diameter of the wire, A is the mean area and n_c is the number of wires in the cable. Four-place logarithms were used for this purpose.

The mean diameter of the wires used in a particular cable will be found to be in close agreement with the empirical formula $d = K \dfrac{D}{N+3}$, D representing the diameter of the cable in inches, N the number of wires in the outer ring of a strand, and K is a constant for any one group of cables of the classification. The value of K is unity for 6 by 19 plow and crucible steel hoisting rope and 6 by 7 guy rope. It is four-fifths for the extra flexible 8 by 19 plow-steel rope, and one-third for the tiller rope. This formula will give the diameter of the wire to within 0.001 or 0.002 inch. For example, in the case of the crucible-cast and plow-steel ropes of three-fourths inch diameters, there are 12 wires in the outer ring of a strand. Accordingly $d = 1.0 \dfrac{0.75}{12+3} = 0.0500$ inch. The mean diameters as given in Table 1 found by individual measurements of all the cables of this group are 0.0509 inch and 0.0503 inch, respectively. Similarly with a 1-inch tiller rope the equation gives $d = \dfrac{1}{3} \dfrac{1.0}{6+3} = 0.0370$ inch, as against 0.0357 from the table.

The mean sectional areas of the cables are given approximately by the formula $A_m = \dfrac{3}{8} D^2$ for the 6 by 19, 8 by 19, and 6 by 7 constructions, and $A_m = \dfrac{1}{4} D^2$ for the 6 by 42 tiller-rope constructions, as may be found by reference to Table 1. The areas A_m may be subject to error of 0.01 or 0.02 square inch. More exact coefficients for D^2 are given in this table and the actual areas, and the range of variation may be noted by comparison with Tables 3 to 12.

TABLE 1.—Mean Diameters of Wires and Sectional Areas of Cables

Diameter of cable, D	Tiller rope, 6 by 42		Guy rope, 6 by 7		Crucible-steel hoisting rope, 6 by 19		Plow-steel hoisting rope, 6 by 19		Plow-steel hoisting rope, 8 by 19	
	Diameter of wires, d	Area of cable, A	Diameter of wires, d	Area of cable, A	Diameter of wires, d	Area of cable, A	Diameter of wires, d	Area of cable, A	Diameter of wires, d	Area of cable, A
¼ inch	0.0090	0.016	0.0264	0.023	0.0164	0.024	0.0167	0.025	0.0139	0.023
⅜ inch	.0140	.039	.0404	.054	.0252	.057	.0255	.058	.0198	.047
½ inch	.0184	.067	.0537	.095	.0338	.102	.0346	.107	.0267	.085
⅝ inch	....	....	.0705	.164	.0423	.160	.0418	.156	.0340	.138
¾ inch	....	....	.0794	.208	.0509	.232	.0503	.226	.0420	.210
⅞ inch	....	....	.0905	.270	.0567	.288	.0603	.325	.0491	.288
1 inch	.0357	.252	.1061	.371	.0657	.386	.0681	.415	.0544	.353
1⅛ inches	....	....	.1205	.479	.0722	.467	....	....	....	....
1¼ inches	....	....	....	....	.0839	.630	.0840	.632	.0678	.549
1½ inches	....	....	....	....	....	....	.1047	.981	.0856	.875

APPROXIMATE FORMULA FOR WIRES

$d = K\dfrac{D}{N+3}$
$\begin{cases} d = \text{diameter of wires.} \\ D = \text{diameter of cable.} \\ N = \text{number of wires in outer ring.} \end{cases}$

$K = 1.0$ for 6 by 19 plow and crucible steel rope and 6 by 7 guy rope.

$K = .8$ for 8 by 19 plow-steel rope.

$K = .33$ for 6 by 42 tiller rope.

APPROXIMATE FORMULA FOR AREAS

$$A = C D^2$$

$A = $ area of cable.

$C = 0.41$ for 6 by 19 plow-steel rope.

$C = .38$ for 6 by 19 crucible-steel rope.

$C = .38$ for 6 by 7 guy rope.

$C = .35$ for 8 by 19 plow-steel rope.

$C = .26$ for 6 by 42 tiller rope.

In view of the time required in making micrometer measurements of the actual wires of a cable and the subsequent somewhat tedious calculations of the aggregate areas of the wires, the above formulas have been found of general utility in giving a quick check on the measurements. They are of value in giving analytical expressions for these functions in mechanical analyses, and in making quick estimates.

(c) *Lays of Strands and Wires.*—The enveloping surface of a strand is a particular case of the tubular surface. This is the surface generated by a sphere of constant radius whose center moves upon a skew curve as directrix. The directrix in the case of a strand of the cable is a helix. The helix is a curve whose tangent makes a constant angle with a fixed straight line.[2] The axis of the central wire of the strand as it winds about the central axis of the cable generates a helix. The distance along the axis of the cable in which this helix makes one complete turn is generally known as the pitch. It is called the lay by wire-rope manufacturers. A wire in a strand winds about the helical axis of the strand as this in turn winds about the axis of the cable. It generates a "compound" helix or a more general form of the simple helix. The distance along the axis of the strand in which the wire makes one complete revolution is called the lay of the wire. The lays of the strands and wires were measured and will presently be discussed.

When the wires twist in the same direction about the axis of the strand as the strand twists about the axis of the cable the construction is known as Lang's lay, and is sometimes called the Albert lay. When, on the other hand, the direction of twist of the wires is in an opposite direction to that of the strands, the construction is known as the regular lay. If the direction of the twist in a strand corresponds to that of a right-handed screw, the wires being twisted in the opposite direction about the axis of the strand, the construction is known as right lay, and vice versa as left lay. There are also right and left lays in the case of the strands of the Lang lay ropes. The right-lay strands are of standard construction.[3] The cables tested are regular right lay throughout, as is shown in Fig. 3.

(d) *Laws of Construction and Formulas for Estimating Purposes.*—From the data of Fig. 1 and Tables 3 to 12 the following laws of construction are deducible with reference to the geo-

---

[2] The general analysis of skew curves is discussed by E. Goursat, Cours d'Analyse (trans. by E. R. Hedrick), chapters 11 and 12.

[3] See American Wire Rope (1913), chapter 3; handbook issued by American Steel & Wire Co.

metric properties of the cables, the diameters of wires, lays, etc. The dimensions are found to be linear functions of the diameter of cable, viz, a constant $\times D$ (very nearly); in other words, if the diameter of wires, of rope core, the lengths of lay, etc., are known for one diameter of cable of the same type of construction, those of another diameter of this type may be found simply by considering the degree of "magnification" in diameters of the second over the first. For example, taking the mean diameter of the wires in a five-eighths inch plow steel of 6 by 19 construction given in Table 1, as 0.0418, the corresponding value for a 1¼-inch diameter cable is 0.0840, approximately twice the other, and so on proportionately for other diameters for this and other specimens.

Similarly the diameters of the rope cores are a certain fraction throughout of the cable diameters. The effective diameter of the rope core is equal to the diameter of the inner cylindrical envelope of strands, and is found by subtracting the diameter of two strands taken along the diameter of cable from that of the cable diameter. It is found in this manner that the effective diameter of the core is one-third the diameter of cable $\dfrac{D}{3}$ for a six-strand cable. It is nearly $\dfrac{D}{2}$ for one of eight strands. For example, for a 1-inch plow-steel rope, there being 5 wires on the diameter of a strand of 0.0681-inch mean diameter, the diameter of strand and rope core is 0.0681 inch by 5 = 0.3405 inch = (approximately) $\dfrac{D}{3}$, and similarly for the plow-steel 8 by 19 construction. For the guy ropes, 3 wires are taken instead, the strand and rope-core diameters being $\dfrac{D}{3}$ as before.

The actual diameters of the rope cores before they enter the cable are approximately $\dfrac{D}{2}$. During fabrication the core is compressed so that the strands bed firmly on the hemp, and the material will fill the grooves formed by the strands and wires. This law holds approximately for the tiller rope, although more difficulty was experienced in measuring the diameters of the strands exactly on account of the presence of the small rope in the strands. The separate strands of tiller rope, however, appear to follow the same laws as the cables.

·The lays of over 150 of the strands were measured for the different diameters of cables of each class. The maximum, minimum, and mean values observed are platted in Fig. 2. The

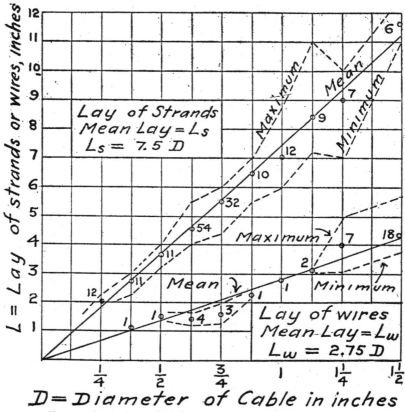

FIG. 2.—*Lays of strands and wires for cables of different diameters*

The lay is the pitch or distance along the axis of cable or strand in which the strand or wire makes a complete revolution. The numerals refer to the number of separate observations taken in determining the mean. The range of variation from the mean is shown by the broken lines.

mean lay of the strands L_s may be taken for purposes of analysis of results as $7.5\ D$. With a 1-inch cable, for example, the strand makes a complete turn around the axis in about $7\frac{1}{2}$ inches. Extreme values as low as 6 inches and as high as $8\frac{1}{2}$ inches were found for this diameter, the practice of the manufacturers varying somewhat in meeting different conditions. More difficulty is experienced in tracing the course of the wires. The mean value for the lay of wires $L_w = 2\frac{3}{4}D$ or $3\ D$ is a fair estimate from the measurements for purposes of analysis. The mean values from the measurements are indicated by a small circle in the figure, the number of observations taken being indicated by the adjoining numerals.

J. B. Smith [4] in discussing English practice some years ago says that "as a general approximation, it may be stated that the lays in strands vary from 2 to 6 inches or about 3 to 4 times the diameter of the rope, while the lays in roping range from about 6 to 12 inches, or 7 to 10 times their diameter. In other words, about 2 to 3 twists are put in the strand to 1 in the rope." The American practice, as indicated in these results, is evidently such as to give a good degree of flexibility of the rope without reducing its efficiency too much in developing the aggregate strength of the wires, the maximum of efficiency being attained with parallel lays of the wire and strands.

The orthogonal projection of the helix formed by the central wire of a strand on a transverse section of the cable is a circle whose diameter is $2/3D$. The corresponding value for the outer wire of a strand referred to the axis of strand is eight-tenths of $1/3D$ for the 6 by 19 construction. Accordingly, taking L_s as the pitch of strand and L_w for the pitch of the outer wire, the relation existing is

$$\frac{\frac{2}{3}D}{L_s} = \frac{\frac{8}{30}D}{L_w} \text{ or } \frac{L_w}{L_s} = 0.4 : \text{i. e., approximately } \frac{2\frac{3}{4}}{7\frac{1}{2}},$$

as found from the mean ratios of the lays as already determined. The angle of slope of the wires referred to the axis of the strand in standard constructions is equal in magnitude, but opposite in direction to the angle of slope of the strand. The effect upon the rope construction is to make the wires on the exposed periphery of cable take an axial direction. The cable by this construction is most effective in developing the highest flexural efficiency of the wires as well as the highest efficiency for abrasive resistance. The axial direction of the wires upon the periphery of cable is well shown in cuts of American cables of the types here considered. It is also carried out in the case of the smaller 6 by 7 strands (considered as units) of the tiller rope, and most other types, except the Lang lay ropes. (See Fig. 3.)

---

[4] Treatise on Wire; Its Manufacture and Usage, 1891.

V. OUTLINE OF METHODS OF TESTS

1. STANDARD LENGTH OF TEST SPECIMENS

In making a tensile test of a cable in such a way as to best approximate actual service conditions, it is desirable as far as practicable to eliminate the local effects of the end connections in the testing machine, even more so than is commonly required in tests of the other materials of construction. If an indefinite length of specimen were possible in making a test, it would undoubtedly give test results more comparable with the conditions of practice. Some engineers, indeed, have advised that the length for tests shall be from 25 to 100 feet. Such lengths are impracticable, not only on account of the additional costs for materials, but also because of the limited heights of the testing machines and the difficulty in handling and preparing specimens. The practical importance of long lengths is believed to be overestimated.

The length chosen for a standard in these tests is 6 feet 8 inches (80 inches). Experience has shown that this length is quite adequate to meet the practical considerations of cost, ease of handling, and the general conditions imposed by the tests. In view of the factors of uncertainty which enter and are incident to the difficulties experienced in rigorously stating the mechanics of a helical strand resting upon a partially elastic rope core, it will be evident that great refinement in this respect is inexpedient.

2. PREPARATION OF CABLES FOR TENSILE TESTS

In making a tensile test it is essential that the specimen shall be free from bends. A flat curvature to the specimen, while insignificant as regards the tensile strength, will effect elongation determinations during the earlier loads appreciably. Such imperfect cables have been discarded in elongation tests. Another important point in preparation of the specimen is that the force shall be applied axially and that there shall be no lost motion due to relative slipping of wires or strands in the sockets. Indeed, if such were the case, there would not be a uniform distribution of the load among the different strands.

Zinc sockets were used in making the tests. The wires slip when babbitted sockets are used at loads as low as 25 per cent of the maximum strength. In preparing the cable for socketing, the ends are first "served" or wound for about 1½ inches with soft wire (one-eighth inch "clothesline" rope was used) at the ends and at a distance from the ends equal to the length of the zinc

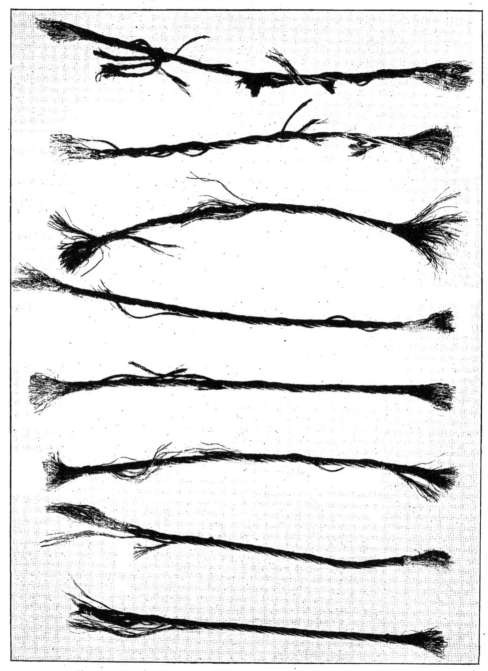

FIG. 3.—*Typical fractures of wire ropes of 1¼ and 1½ inches diameters*

The zinc has been melted from the end sockets to show "brooming" of wires in preparing a test specimen socket. In the present tests the length varied from 5 to 9 inches in a range of diameters of one-fourth to 1½ inches. Special attachments were used on the few cables over 1½ inches in diameter tested in the Emery machine. (See note 5, p. 20.)

After the cables were served as described, they were slipped through wedge-shaped cast-steel blocks, which acted later not only as molds for the zinc surrounding the unraveled wires and forming the sockets but also as pulling blocks when these were inserted in the wedge-shaped opening in the heads of the testing machine. (See Fig. 6.) Solid blocks were used on the cables above seven-eighths inch in diameter. Split blocks were used for convenience on the smaller cables.

After the blocks were placed on a cable, the specimen was clamped in a vise, the serving wire was removed at the ends, and the strands and wires were opened or frayed out as far as the second serving. The rope core was removed in a distance equal to the length of the socket. The individual wires were thoroughly cleaned with waste and the free use of gasoline and were then carefully wiped to insure their freedom from the least amount of oil which might occasion a slipping of the wires in the zinc. In some cases the frayed portions were dipped in a pail of caustic-soda solution. This is not necessary when the lubricant is carefully removed and the wires are thoroughly cleaned. A large number of the wires are often bent back on themselves at the ends for an inch or two to insure a good bond in the zinc. This was not done on the cables of Fig. 3.

Commercial spelter was heated in the small crucible furnace shown in Fig. 4, the temperature being sufficient to give a good degree of fluidity. It must not be too hot; otherwise the outer wires in contact with the molten spelter have their strength impaired by the heat. This makes it difficult in securing the best results in testing the one-fourth inch diameter specimens on account

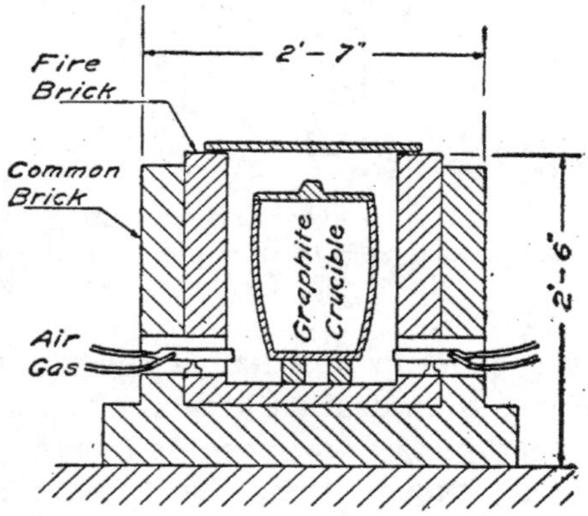

Fig. 4.—*Furnace and crucible for melting zinc for cable sockets*

of the annealing effect on the small wires. The number of fractures at the sockets is usually a maximum for small-diameter cables.

In pouring the molten zinc into the cone-shaped cavity of the block containing the frayed-out wires, a special alignment frame, as shown in Fig. 5, was used. The cable was made truly axial with the testing blocks, a small ring of fire clay was added to prevent seepage of molten metal at the base of block, and the zinc was ladled as rapidly as possible to insure a uniformly cast conical socket. The frayed-out portions, after zinc has been melted off, is well shown in Fig. 3.

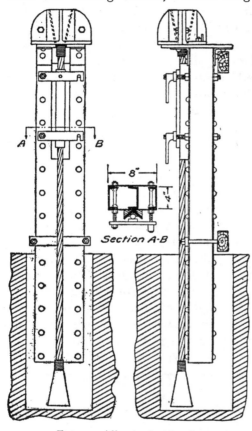

FIG. 5.—*Alignment apparatus*

The specimen is clamped as shown and molten zinc poured into mold at upper end. The finished socket is shown at lower end

3. METHODS OF TESTING

Most of the smaller-size cables of diameters from one-fourth to seven-eighths inch, inclusive, were tested in a 100 000-pound Olsen machine. A few of these were tested in a 600 000-pound Olsen machine, together with the remaining cables varying in diameters from 1 to 1½ inches, inclusive. The cables of diameters greater than 1½ inches were tested in the 1 150 000-pound Emery machine of the Bureau at Washington.[5]

A diagrammatic sketch of a cable in position in the 600 000-pound machine is given in Fig. 6 to indicate the methods pursued in making tests. This is also typical of the method used with the 100 000-pound machine, except that the split blocks were used, as previously described, for convenience in handling and inserting the specimens in the machine.

Power was applied at the slower speeds during the earlier loadings. This gives an opportunity for the strands and wires to properly bed upon one another during the application of the

---

[5] A description of the methods of preparation and tests of the cables of 2 to 3¼-inch diameters is given in Engineering Record, 74, p. 81; July 15, 1916.

loads. The speed was decreased a little after the earlier loadings.
The power was then removed, when strain measurements were
taken. Proximity to the breaking load was usually indicated

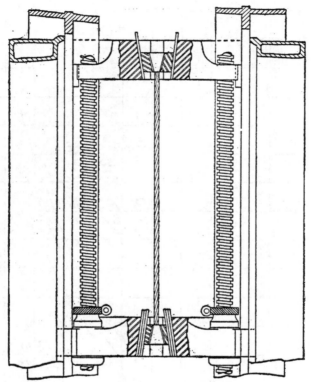

FIG. 6.—*Sectional view showing arrangement of specimen in
the testing machine*

slightly in advance by the snapping of a few of the interior wires,
which were accompanied by sharp metallic reports. This was
soon followed by the fracture of several strands of a specimen.
Characteristic fractures are shown in the group of 1¼ by 6 by 19
plow-steel cables, shown in Fig. 3.

VI. DISCUSSION OF THE RESULTS OF TENSILE TESTS OF CABLES

1. ANALYSIS OF OBSERVED MAXIMUM LOADS

The maximum loads recorded on the beam of the testing
machine are given for each cable in Tables 3 to 13. The arithmetical means of these loads for each group of specimens are
shown in the tables, and have been platted in Fig. 7 as functions
of the diameters of the cables.

The maximum loads are quadratic functions of the diameters, and the relations which exist may be expressed by simple empirical equations of the form $L = CSD^2$, where L represents the

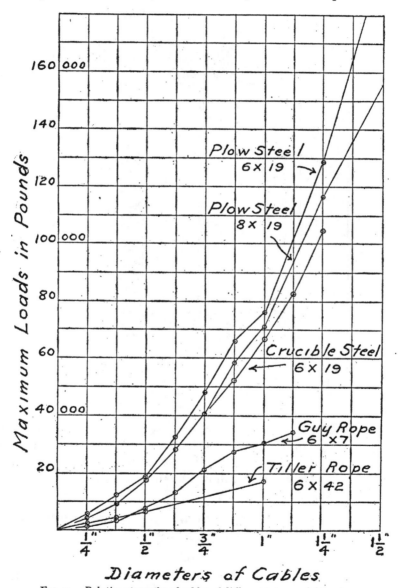

FIG. 7.—*Relative strengths of cables of different types and diameters*

The values indicated by the small circles are the averages of results given in Tables 3 to 12, inclusive

observed maximum load, S = the load which a 6 by 19 plow steel of 1 inch in diameter will sustain, D is the diameter of the cables, and C is a parameter varying for the groups, but nearly constant for any one group. Let it be conceived that the loads from each individual test for the 6 by 19 plow-steel group are platted as functions of the diameters after the manner of Fig. 7, but all the observations being included. The mean curves already shown in the figure will trace a central path through the zone comprising the observations. The lower frontier of this field is defined by the minimum results recorded for each test, and may be analytically expressed by the equation $L = C$ 75 000 D^2. The parameter C will vary from 0.9 to 1.1, and has a mean value of approximately unity.

If the other groups are similarly platted, the lower frontiers of the 8 by 19 plow steel and the 6 by 19 crucible cast-steel groups will be expressed fairly well by the same equation, but C varies from 0.80 to 1.00, with a mean value of about 0.85. In the case of the guy and tiller ropes C varies from 0.3 to 0.45, with a mean value of approximately 0.35.

These equations show that the proportionate minimum strengths of the different groups are approximately in the ratios of 10:8½:8½:3½:3½. The probable load which a cable will carry, as expressed by Fig. 7, will be about 5 to 12 per cent higher than the minimum values recorded; in other words, the values C will need to be increased approximately these amounts.

The strengths called for in the specifications and the 1910 standard strengths of the manufacturers agree quite closely, as a rule, with the minimum values observed, which define the lower frontiers of the groups. Accordingly, if it is desired to insure that the maximum load a cable of the classes given sustains shall not fall below a certain limit, that limit is expressed fairly well by the Isthmian Canal specifications or the standard strengths of the manufacturers. If, on the other hand, it is desired to obtain an estimate of the probable load that cables of this classification will carry, it will usually be somewhat in excess of the standard strengths, as a rule, say, about 10 per cent. In other words, the standard strengths are conservative and cover the standard types of steel. The higher mean strengths of the cables are influenced partly by the fact that improved steels have been used by certain manufacturers in several cases in meeting the provisions of the specifications. The maximum loads above the means may indicate the presence of superior plow and crucible steels, or they may be fortuitous, simply high values for the standard steels.

The equations

$$L = C\ 75\ 000\ D^2:\ C\begin{cases} = 0.9 \text{ to } 1.1;\ 6 \text{ by } 19 \text{ plow steel} \\ = 0.8 \text{ to } 1.0;\ 6 \text{ by } 19 \text{ crucible, } 8 \text{ by } 19 \text{ plow} \\ = 0.3 \text{ to } 0.45;\ 6 \text{ by } 42 \text{ tiller, } 6 \text{ by } 7 \text{ guy} \end{cases}$$

should be considered to have the limitations of empirical formulæ, but they are useful in expressing the test results of a large amount of experimental data in a relatively small compass for the approximate general estimation and designing purposes of engineers. Table 2, following, shows the relation of the loads calculated by

TABLE 2.—Relation of Observed Breaking Loads of 6 by 19 Plow-Steel Cables to 1910 Standard Strengths and the Formula $L=C\ 75\ 000\ D^2$

Diameter, in inches, D	Standard strengths=Isthmian Canal specifications, in pounds	Formula $L=C\ 75\ 000\ D^2$ $C=1$	Observed breaking loads from tests		
			First and second minimums	Maximums	Mean
¼	5300	4680	5250 5610	5970	5610
⅜	11 500	10 550	10 600 12 150	13 000	12 140
½	20 000	18 750	17 900 17 930	20 600	18 680
⅝	31 000	29 300	29 550 29 940	35 990	32 760
¾	46 000	42 200	43 500 44 210	52 620	47 920
⅞	58 000	57 400	56 570 58 650	72 300	65 800
1	76 000	75 000	75 710 76 270	76 270	76 000
1⅛	94 000	94 900	....	....	....
1¼	116 000	117 000	108 000 119 000	164 800	128 800
1⅜	144 000	142 000	....	....	....
1½	164 000	168 750	148 000 163 500	233 280	193 940

the formula to the standard strengths and the results of tests in the case of the 6 by 19 plow-steel cables as given in Tables 3 to 12. The first figure in the column of "minimums" gives the lowest breaking loads observed, while the second figure records the next to the lowest loads. The values as given by the formula agree quite closely with the standard strengths and the lowest breaking loads recorded for the tests. The minimum breaking loads recorded for the 1¼ and 1½ inch diameters are believed to be abnormally low. The second figures in the scale of observed values are believed to be more representative of the minimum strengths of these cables and are in closer agreement with the standard strengths and the specifications.

TABLE 3.—Tensile Strengths of ¼-Inch Diameter Steel Cables

Serial No.	General data						Observed mechanical data			
	Manufacturer	Type of steel	Use in practice	Strands and wires	Diameter of rope core	Sectional area observed	Maximum load		Number of strands broken	
									In body	At socket
					Inch	Inch²	Pounds	Lbs./in.²		
1....	M-4..	Swedish iron..	Light hoist....	6 by 19	⅛	0.0174	2920	167 800	......	4
2....	M-4..	Galvanized steel	Guys, rigging..	6 by 7	⅛	.0206	1400	67 970	4	......
3....	M-4..	.....do.........	.....do.........	6 by 7	⅛	.0259	1880	72 700	4	......
Mean	......	......	......	......	......	.023	1640	70 335	......	......
4....	M-4..	Swedish iron..	Boat tillers, etc.	6 by 42	3/32	.0160	2240	139 800	......	6
5....	M-4..	.....do.........	.....do.........	6 by 42	3/32	.0179	2340	130 700	4	......
6....	M-4..	.....do.........	.....do.........	6 by 42	3/32	.0160	2440	152 500	......	4
7....	M-4..	.....do.........	.....do.........	6 by 42	3/32	.0141	2150	152 470	......	4
Mean	......	......	......	......	......	.016	2293	143 868	......	......
8....	M-4..	Crucible cast steel	Light hoist....	6 by 19	⅛	.0232	4650	200 400	3	......
9....	M-4..	.....do.........	.....do.........	6 by 19	⅛	.0228	4430	194 300	2	......
10....	M-4..	.....do.........	.....do.........	6 by 19	⅛	.0229	4490	195 100	......	4
11....	M-4..	.....do.........	.....do.........	6 by 19	⅛	.0229	4350	190 000	......	3
12....	M-4..	.....do.........	.....do.........	6 by 19	⅛	.0229	4200	183 400	......	3
13....	M-4..	.....do.........	.....do.........	6 by 19	⅛	.0259	5230	202 000	1	......
14....	M-4..	.....do.........	.....do.........	6 by 19	⅛	.0259	5610	216 500	......	1
15....	M-4..	.....do.........	.....do.........	6 by 19	⅛	.0259	5350	206 600	......	6
Mean	......	......	......	......	......	.024	4790	198 663	......	......
16....	M-10.	Plow steel.....	Light hoist....	6 by 19	⅛	.0260	5610	215 800	......	3
17....	M-10.	.....do.........	.....do.........	6 by 19	⅛	.0260	5970	229 600	......	6
18....	M-9..	.....do.........	.....do.........	6 by 19	⅛	.0229	5250	229 300	......	2
Mean	......	......	......	......	......	.025	5610	224 900	......	......
19....	M-11.	Plow steel.....	Extra flexible.	8 by 19	⅛	.0270	4700	174 000	......	......
20....	M-9..	.....do.........	.....do.........	8 by 19	⅛	.0203	4800	236 500	......	4
21....	M-9..	.....do.........	.....do.........	8 by 19	⅛	.0202	5780	286 200	......	7
Mean	......	......	......	......	......	.023	5093	232 233	......	......

More exact equations may be derived which will fit the results of the observations very closely; but they lack the simplicity of form of the expressions which have been given, and little is to be gained by exact expressions when the relatively large variations which occur in tests of this nature are considered.

TABLE 4.—Tensile Strengths of ⅜-Inch Diameter Steel Cables

Serial No.	General data						Observed mechanical data			
	Manufacturer	Type of steel	Use in practice	Strands and wires	Diameter of rope core	Sectional area observed	Maximum load		Number of strands broken	
									In body	At socket
					Inch	Inch²	Pounds	Lbs./in.²		
22	M-4	Swedish iron	Light hoist	6 by 19	$\frac{7}{16}$	0.0605	4585	75 900	3	......
23	M-4	.....do	Boat tillers, etc.	6 by 42	$\frac{7}{16}$	.0388	4460	115 000	......	2
24	M-4	.....do	.....do	6 by 42	$\frac{7}{16}$	.0388	4680	120 620	3	......
25	M-4	.....do	.....do	6 by 42	$\frac{7}{16}$	.0388	4140	106 700	......	
26	M-4	.....do	.....do	6 by 42	$\frac{7}{16}$	.0388	4090	105 410	......	
Mean						.039	4343	111 933	......	
27	M-4	Galvanized steel	Rigging guys	6 by 7	$\frac{7}{16}$	.0556	3920	70 500	2	......
28	M-4	.....do	.....do	6 by 7	$\frac{7}{16}$	.0528	3920	74 200	2	......
29	M-4	.....do	.....do	6 by 7	$\frac{7}{16}$	.0552	3990	72 200	3	......
30	M-4	.....do	.....do	6 by 7	$\frac{7}{16}$	.0552	4010	72 600	......	1
31	M-4	.....do	.....do	6 by 7	$\frac{7}{16}$	.0526	4020	76 400	2	......
32	M-4	.....do	.....do	6 by 7	$\frac{7}{16}$	.0526	3780	71 900	......	2
33	M-4	.....do	.....do	6 by 7	$\frac{7}{16}$	.0526	3860	73 400	3	......
Mean						.054	3929	73 029	......	
34	M-4	Crucible cast steel	Hoisting	6 by 19	¼	.0563	10 270	182 400	......	4
35	M-4	.....do	.....do	6 by 19	¼	.0567	10 320	182 000	......	3
Mean						.057	10 295	182 200	......	
36	M-4	Plow steel	Hoisting	6 by 19	$\frac{7}{16}$	.0654	12 800	195 800	......	6
37	M-4	.....do	.....do	6 by 19	$\frac{7}{16}$	.0544	10 600	194 900	......	4
38	M-4	.....do	.....do	6 by 19	$\frac{7}{16}$	.0544	13 000	239 000	......	6
39	M-4	.....do	.....do	6 by 19	$\frac{7}{16}$	.0577	12 150	210 570	......	4
Mean						.058	12 138	210 118	......	
40	M-11	Plow steel	Extra flexible hoist	8 by 19	$\frac{7}{16}$	.0477	9720	203 500	......	
41	M-2	.....do	.....do	8 by 19	$\frac{7}{16}$	.0386	8800	228 000	......	7
42	M-2	.....do	.....do	8 by 19	$\frac{7}{16}$	.0508	8510	167 500	4	......
43	M-9	.....do	.....do	8 by 19	$\frac{7}{16}$	.0478	10 680	223 400	......	2
44	M-10	.....do	.....do	8 by 19	$\frac{7}{16}$	.0477	9600	201 300	6	......
45	M-10	.....do	.....do	8 by 19	$\frac{7}{16}$	.0477	9700	203 400	4	......
Mean						0.047	9502	204 517	......	

TABLE 5.—Tensile Strengths of ½-Inch Diameter Steel Cables

Serial No.		General data					Observed mechanical data			
	Manu-fac-turer	Type of steel	Use in practice	Strands and wires	Diam-eter of rope core	Sec-tional area ob-served	Maximum load		Number of strands broken	
									In body	At sock-et
					Inch	Inch ^{2}	Pounds	Lbs./in.^{2}		
46....	M-4..	Swedish iron..	Boat tillers, etc.	6 by 42	$\frac{1}{16}$	0.0781	6690	85 600	......	5
47....	M-4..	.....do.........	.....do.........	6 by 42	$\frac{1}{16}$	.0640	6100	95 300	......	3
48....	M-4..	.....do.........	.....do.........	6 by 42	$\frac{1}{16}$	.0714	5850	81 900	......	3
49....	M-4..	.....do.........	.....do.........	6 by 42	$\frac{1}{16}$	.0597	6450	108 100	3	......
50....	M-4..	.....do.........	.....do.........	6 by 42	$\frac{1}{16}$	.0641	6780	105 800	1	......
Mean						.067	6374	95 280		
51....	M-4..	Galvanized steel.	Rigging and guys.	6 by 7	¼	.1010	7940	78 500	......	2
52....	M-4..	.....do.........	.....do.........	6 by 7	¼	.0892	8000	89 700	1	......
53....	M-4..	.....do.........	.....do.........	6 by 7	¼	.0892	6930	77 700	3	......
54....	M-4..	.....do.........	.....do.........	6 by 7	¼	.0998	7570	75 950	3	......
Mean						.095	7610	80 463		
55....	M-4..	Crucible-cast steel.	Hoist.........	6 by 19	¼	.1050	18 280	174 100	2	......
56....	M-4..	.....do.........	.....do.........	6 by 19	¼	.1035	16 280	157 300	......	1
57....	M-4..	.....do.........	.....do.........	6 by 19	¼	.0971	18 960	195 300	......	5
Mean						.102	17 840	175 566		
58....	M-9..	Plow steel.....	.....do.........	6 by 19	$\frac{3}{16}$	.1070	18 340	171 300	4	......
59....	M-9..	.....do.........	.....do.........	6 by 19	$\frac{3}{16}$	.1070	17 930	167 500	2	......
60....	M-9..	.....do.........	.....do.........	6 by 19	$\frac{3}{16}$	.1070	17 900	167 200	2	......
61....	M-9..	.....do.........	.....do.........	6 by 19	$\frac{3}{16}$	.1049	18 600	177 300	2	......
62....	M-9..	.....do.........	.....do.........	6 by 19	$\frac{3}{16}$	.1076	20 600	191 500	......	2
Mean						.107	18 674	174 960		
63....	M-11.	.....do.........	Extra-flexible hoisting.	8 by 19	$\frac{1}{16}$	.0869	18 320	211 000	......	......
64....	M-11.	.....do.........	.....do.........	8 by 19	$\frac{1}{16}$	.0806	19 980	248 000	7	......
65....	M-9..	.....do.........	.....do.........	8 by 19	$\frac{1}{16}$	.0882	16 550	187 600	2	......
Mean						.085	18 283	215 530		

TABLE 6.—Tensile Strengths of ⅝-Inch Diameter Steel Cables

Serial No.	Manufacturer	Type of steel	Use in practice	Strands and wires	Diameter of rope core	Sectional area observed	Maximum load		Number of strands broken	
							Pounds	Lbs./in.²	In body	At socket
					Inch	Inch²	Pounds	Lbs./in.²		
66....	M-4..	Galvanized steel.	Rigging and guys.	6 by 7	¼	0.1664	13 430	80 700	......	4
67....	M-4..	.....do..........	.....do.........	6 by 7	¼	.1570	12 950	82 500	2	......
68....	M-4..	.....do..........	.....do.........	6 by 7	¼	.1616	12 430	76 900	3	......
69....	M-4..	.....do..........	.....do.........	6 by 7	¼	.1722	13 600	78 980	3	......
Mean	......				......	.164	13 103	79 770	......	......
70....	M-4..	Crucible-cast steel.	Hoisting rope..	6 by 19	7/16	.1466	27 170	185 300	......	2
71....	M-4..	.....do..........	.....do.........	6 by 19	7/16	.1550	26 220	169 000	......	1
72....	M-4..	.....do..........	.....do.........	6 by 19	7/16	.1570	26 860	171 000	3	......
73....	M-4..	.....do..........	.....do.........	6 by 19 by 6	7/16	.1648	26 830	162 800	2	......
74....	M-4..	.....do..........	.....do.........	6 by 19 by 6	7/16	.1648	27 260	165 400	1	......
75....	M-4..	.....do..........	.....do.........	6 by 19	7/16	.1626	29 150	179 300	2	......
76....	M-4..	.....do..........	.....do.........	6 by 19 by 6	7/16	.1651	32 020	194 000	3	......
77....	M-4..	.....do..........	.....do.........	6 by 19 by 6	7/16	.1651	30 930	187 400	4	......
Mean	......				......	.160	28 305	176 775	......	......
78....	M-1..	Plow steel.....	.....do.........	6 by 19	7/16	.1505	31 200	207 300	3	......
79....	M-1..	.....do..........	.....do.........	6 by 19	7/16	.1433	31 830	222 100	......	1
80....	M-1..	.....do..........	.....do.........	6 by 19	7/16	.1433	33 510	233 800	3	......
81....	M-2..	.....do..........	.....do.........	6 by 19 by 6	7/16	.1444	33 940	235 000	1	......
82....	M-2..	.....do..........	.....do.........	6 by 19	7/16	.1469	35 990	245 000	......	1
83....	M-4..	.....do..........	.....do.........	6 by 19 by 6	7/16	.1677	32 250	192 300	......	a 6
84....	M-4..	.....do..........	.....do.........	6 by 19 by 6	7/16	.1677	33 290	198 500	......	4
85....	M-9..	.....do..........	.....do.........	6 by 19	7/16	.1521	29 940	196 900	2	......
86....	M-8..	.....do..........	.....do.........	- 6 by 19	7/16	.1621	34 000	209 400	......	a 6
87....	M-2..	.....do..........	.....do.........	6 by 19	7/16	.1579	33 260	210 600	3	......
88....	M-1..	.....do..........	.....do.........	6 by 19	7/16	.1433	33 970	237 100	......	4
89....	M-2..	.....do..........	.....do.........	6 by 19	7/16	.1579	32 280	204 400	......	5
90....	M-2..	.....do..........	.....do.........	6 by 19	7/16	.1579	30 350	192 200	1	......
91....	M-2..	.....do..........	.....do.........	6 by 19	7/16	.1579	31 440	199 100	4	......
92....	M-2..	.....do..........	.....do.........	6 by 19	7/16	.1579	31 390	198 800	3	......
93....	M-2..	.....do..........	.....do.........	6 by 19	7/16	.1579	32 100	203 300	3	......
94....	M-9..	.....do..........	.....do.........	6 by 19	7/16	.1592	30 200	189 700	......	3
95....	M-5..	.....do..........	.....do.........	6 by 19	7/16	.1485	31 260	210 500	......	2
96....	M-5..	.....do..........	.....do.........	6 by 19	7/16	.1485	29 550	199 000	......	3
97....	M-10.	.....do..........	.....do.........	6 by 19	7/16	.1690	35 960	212 800	4	......
98....	M-10.	.....do..........	.....do.........	6 by 19	7/16	.1700	35 900	211 200	4	......
99....	M-10.	.....do..........	.....do.........	6 by 19	7/16	.1601	34 720	216 900	......	3
100...	M-10.	.....do..........	.....:do.........	6 by 19	7/16	.1601	35 230	220 000	3	......
Mean	......				......	.156	32 763	210 691	......	......

a Parts of each strand unbroken.

TABLE 6.—Tensile Strengths ⅝-Inch Diameter Steel Cables—Continued

Serial No.	Manu-fat-turer	Type of steel	Use in practice	Strands and wires	Diam-eter of rope core	Sec-tional area ob-served	Maximum load		Number of strands broken	
									In body	At socket
					Inch	Inch²	Pounds	Lbs./in.²		
101...	M-9..	Galvanized plow steel.	Strong guy ropes, etc.	6 by 19	¼	.1590	32 920	207 000	3	......
102...	M-9..	.....do.........	.....do.........	6 by 19	¼	.1510	36 300	240 400	1	......
103...	M-9..	.....do.........	.....do.........	6 by 19	¼	.1510	36 700	243 100	......	2
104...	M-2..	.....do.........	.....do.........	6 by 19	¼	.1432	30 480	212 900	3	......
Mean						.151	34 100	225 850	......	......
105...	M-4..	Plow steel.....	Extra-flexible hoisting rope.	8 by 19	⅜	.1280	24 980	195 200	......	1
106...	M-4..	.....do.........	.....do.........	8 by 19	⅜	.1070	29 640	277 000	......	4
107...	M-9..	.....do.........	.....do.........	8 by 19	⅜	.1398	30 900	221 000	......	2
108...	M-4..	.....do.........	.....do.........	8 by 19	⅜	.1502	30 680	204 300	......	5
109...	M-4..	.....do.........	.....do.........	8 by 19	⅜	.1381	30 600	221 600	4	......
110...	M-4..	.....do.........	.....do.........	8 by 19	⅜	.1450	29 800	204 100	......	4
111...	M-4..	.....do.........	.....do.........	8 by 19	⅜	.1460	29 710	203 500	4	......
112...	M-4..	.....do.........	.....do.........	8 by 19	⅜	.1240	24 660	198 900	......	4
113...	M-4..	.....do.........	.....do.........	8 by 19	⅜	.1208	25 300	209 400	5·	......
114...	M-4..	.....do.........	.....do.........	8 by 19 by 6	⅜	.1374	25 890	188 400	......	5
115...	M-9..	.....do.........	.....do.........	8 by 19	⅜	.1460	27 250	186 800	2	......
116...	M-9..	.....do.........	.....do.........	8 by 19	⅜	.1460	27 090	185 500	4	......
117...	M-10.	.....do.........	.....do.........	8 by 19	⅜	.1500	29 030	193 530	4	......
118...	M-10.	.....do.........	.....do.........	8 by 19	⅜	.1500	29 320	195 470	4	......
119...	M-10.	.....do.........	.....do.........	8 by 19	⅜	.1441	28 150	195 300	2	......
120...	M-10.	.....do.........	.....do.........	8 by 19	⅜	.1441	27 480	190 700	3	......
121...	M-9..	.....do.........	.....do.........	8 by 19	⅜	.1317	27 500	208 780	......	......
Mean						.138	28 116	204 675	......	......

2. OBSERVED MAXIMUM STRESSES DISCUSSED

The strengths of the cables may be placed on a more appropriate basis for comparison with the strengths of their constituent wires by considering the stresses which were developed. For this purpose the maximum loads were reduced to stresses by dividing them by the aggregate cross-sectional areas of the wires calculated from micrometer measurements of the different diameters of the wires in each cable as has been described. The results are given in Tables 3 to 13 under the heading "Maximum load, pounds per square inch," the arithmetical means being recorded for the different classes. It will be found that the mean value of the maximum loads for a group of cables having the same diameter, when divided by the mean cross-sectional area of that group, will agree fairly closely with the mean of the stresses figured for each individual specimen. In lieu of precise knowledge as to the cross-sectional areas of the cables given in the tables of standard strengths by the manufacturers, the mean areas given in Tables 3 to 13 will be used. They have been regrouped for reference in Table 1, already given.

TABLE 7.—Tensile Strengths of ¾-Inch Diameter Steel Cables

Serial No.	General data						Observed mechanical data			
	Manufacturer	Type of steel	Use in practice	Strands and wires	Diameter of rope core	Sectional area observed	Maximum load		Number of strands broken	
									In body	At socket
					Inch	Inch [2]	Pounds	Lbs./in.[2]		
122...	M-4..	Galvanized steel.	Guys..........	6 by 7	5/16	0.1905	14 190	74 500	......	3
123...	M-4..	.....do.........	.....do.........	6 by 7	5/16	.2160	24 500	113 430	1	......
124...	M-4..	.....do.........	.....do.........	6 by 7	5/16	.2160	24 720	114 440	2	......
Mean	......	......	......	......	......	.208	21 137	100 790	......	......
125...	M-4..	Crucible cast steel.	Hoisting.......	6 by 19	⅜	.2173	39 000	179 500	3	......
126...	M-4..	.....do.........	.....do.........	6 by 19	⅜	.2130	36 610	171 900	......	......
127...	M-4..	.....do.........	.....do.........	6 by 19	⅜	.2130	37 220	174 700	1	......
128...	M-4..	.....do.........	.....do.........	6 by 19 by 6	⅜	.2486	44 020	177 100	2	3
129...	M-4..	.....do.........	.....do.........	6 by 19 by 6	⅜	.2400	41 680	173 700	......	3
130...	M-4..	.....do.........	.....do.........	6 by 19 by 6	⅜	.2400	41 880	174 500	3	......
131...	M-4..	.....do.........	.....do.........	6 by 19 by 6	⅜	.2405	43 040	179 000	3	......
132...	M-4..	.....do.........	.....do.........	6 by 19 by 6	⅜	.2376	42 600	179 300	4	......
133...	M-4..	.....do.........	.....do.........	6 by 19 by 6	⅜	.2351	40 560	172 500	2	......
134...	M-4..	.....do.........	.....do.........	6 by 19 by 6	⅜	.2351	42 450	180 600	3	......
Mean	......	......	......	......	......	.232	40 906	176 280	......	......
135...	M-4..	Plow steel.....	Hoisting.......	6 by 19	⅜	.2350	48 030	204 400	4	......
136...	M-4..	.....do.........	.....do.........	6 by 19	⅜	.2180	43 500	199 400	......	3
137...	M-4..	.....do.........	.....do.........	6 by 19	⅜	.2155	49 780	231 000	4	......
138...	M-2..	.....do.........	.....do.........	6 by 19 by 6	⅜	.2176	46 210	212 400	3	......
139...	M-1..	.....do.........	.....do.........	6 by 19	⅜	.2063	45 820	222 100	......	2
140...	M-1..	.....do.........	.....do.........	6 by 19	⅜	.2329	52 620	225 900	2	......
141...	M-4..	.....do.........	.....do.........	6 by 19 by 6	⅜	.2387	46 640	195 400	......	1
142...	M-9..	.....do.........	.....do.........	6 by 19	⅜	.2174	44 210	203 400	......	4
143...	M-8..	.....do.........	.....do.........	6 by 19	⅜	.2315	46 440	200 600	......	1
144...	M-10.	.....do.........	.....do.........	6 by 19	⅜	.2322	49 300	212 300	3	......
145...	M-10.	.....do.........	.....do.........	6 by 19	⅜	.2322	48 430	208 700	......	1
146...	M-10.	.....do.........	.....do.........	6 by 19	⅜	.2322	45 690	196 700	......	3
147...	M-9..	.....do.........	.....do.........	6 by 19	⅜	.2262	51 600	228 100	......	3
148...	M-9..	.....do.........	.....do.........	6 by 19	⅜	.2310	52 550	227 500	3	......
Mean	......	......	......	......	......	.226	47 916	211 993	......	......

TABLE 7.—Tensile Strengths of ¾-Inch Diameter Steel Cables—Continued

Serial	Manufacturer	Type of steel	Use in practice	Strands and wires	Diameter of rope core	Sectional area observed	Maximum load		Number of strands broken	
									In body	At socket
					Inch	Inch²	Pounds	Lbs./in.²		
149...	M-4..	Plow steel....	Extra flexible hoisting cable.	8 by 19	⅜	0.1780	38 530	216 500	......	4
150...	M-9..	......do.........	.....do.........	8 by 19	⅜	.2065	42 730	206 900	4	......
151...	M-4..	......do.........	.....do.........	8 by 19	⅜	.2047	37 560	183 500	2	......
152...	M-4..	......do.........	.....do.........	8 by 19	⅜	.2047	36 900	180 300	3	......
153...	M-4..	......do.........	.....do.........	8 by 19 by 6	⅜	.2224	45 300	203 500	2	......
154...	M-4..	......do.........	.....do.........	8 by 19 by 6	⅜	.2224	44 290	198 900	4	......
155...	M-4..	......do.........	.....do.........	8 by 19 by 6	⅜	.1988	36 330	182 800	......	3
156...	M-4..	......do.........	.....do.........	8 by 19 by 6	⅜	.2215	41 950	189 400	4	......
157...	M-9..	......do.........	.....do.........	8 by 19	⅜	.2200	36 900	167 700	1	......
158...	M-9..	......do.........	.....do.........	8 by 19	⅜	.2200	36 495	165 800	3	......
159...	M-10.	......do.........	.....do.........	8 by 19	⅜	.2078	39 630	190 700	......	1
160...	M-10.	......do.........	.....do.........	8 by 19	⅜	.2078	39 830	191 700	2	......
161...	M-9..	......do.........	.....do.........	8 by 19	⅜	.2120	44 270	208 800	......	5
162...	M-10.	......do.........	.....do.........	8 by 19	⅜	.2104	38 900	184 890	1	......
163...	M-10.	......do.........	.....do.........	8 by 19	⅜	.2104	41 720	198 290	3	......
164...	M-10.	......do.........	.....do.........	8 by 19	⅜	.2104	42 580	202 380	......	2
165...	M-10.	......do.........	.....do.........	8 by 19	⅜	.2104	41 560	197 530	3	......
Mean						.210	40 322	192 329		
166...	M-4..	Plow steel......	Steam shovel, etc.	6 by 37	⅜	.2080	48 450	232 900	......	4
167...	M-6..	......do.........	......do.........	6 by 37	⅜	.2350	37 720	160 500	......	3
Mean						.221	43 085	196 700		

The observed maximum stresses for each class of cables have been platted in the upper curves of Figures 8, 9, and 10. The lower curves give the standard strengths of the manufacturers (or those of the specifications), divided by the mean sectional areas as given. It will be seen that there is, in general, an approximate parallelism of the two sets of curves. This indicates that there is a certain correspondence between the tests described and those made by the manufacturers' committee. The depressions and ridges of one set of curves, for example, generally correspond with those of the other. The fact that the curve for the tests of the 1-inch tiller ropes falls below the curve of the manufacturers is doubtless explained by the fact that only two cables were available to the experimenters for tests, and the mean value platted is probably not truly representative.

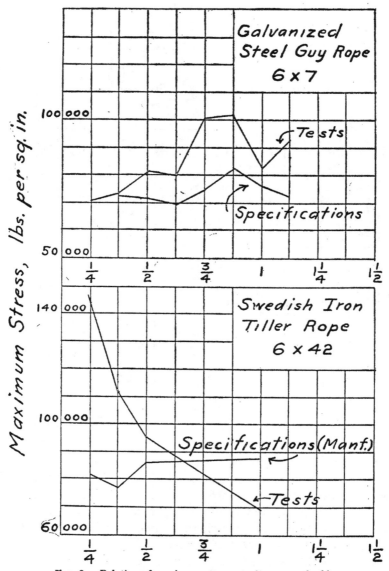

Fig. 8.—*Relation of maximum stresses to diameters of cables*

The upper curves are plats of the averages given in Tables 3 to 12, inclusive. The lower curve for guy rope gives the minimum strengths required by the Isthmian Canal Commission's specifications of 1912, these being slightly lower than the manufacturer's standard strengths of 1910 for iron guy rope. The lower curve for the tiller rope gives the strengths mentioned by one of the manufacturers for iron tiller rope, the minimum strengths of the Canal Commission not being specified

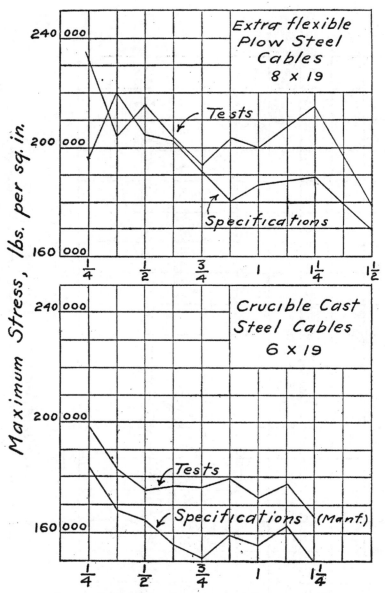

FIG. 9.—*Relation of maximum stresses to diameters of cables*

The upper curves are plats of the averages given in Tables 3 to 12, inclusive. The lower curve of the plow-steel cables gives the minimum strengths required by the Isthmian Canal Commission's specifications of 1912, these being identical with the manufacturers' standard strengths of 1910. The lower curve of the crucible cast-steel cables gives the manufacturers' standard strengths of 1910, the Canal Commission's specifications not calling for this steel

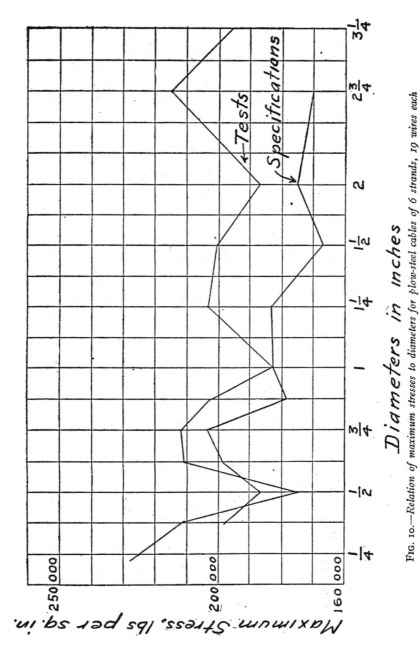

Fig. 10.—*Relation of maximum stresses to diameters for plow-steel cables of 6 strands, 19 wires each*

The upper curve is a plat of the averages given in Tables 3 to 13, inclusive. The lower curve gives the minimum strengths required by the Isthmian Canal Commission's specifications of 1912, these being identical with the manufacturers' standard strengths of 1910

TABLE 8.—Tensile Strengths of ⅞-Inch Diameter Steel Cables

Serial No.	Manufacturer	Type of steel	Use in practice	Strands and wires	Diameter of rope core	Sectional area observed	Maximum load		Number of strands broken	
									In body	At socket
					Inch	Inch ^{2}	Pounds	Lbs./in.^{2}		
168...	M-4..	Galvanized steel.	Rigging and guys.	6 by 7	$\frac{7}{16}$	0. 2640	22 410	84 900	......	2
169...	M-4..	.....do.........	.....do.........	6 by 7	$\frac{7}{16}$	. 2730	30 230	110 730	......	1
170...	M-4..	.....do.........	.....do.........	6 by 7	$\frac{7}{16}$	. 2730	29 830	109 270	3	......
Mean						. 270	27 490	101 633		
171...	M-4..	Crucible - cast steel.	Hoisting.......	6 by 19 by 6	$\frac{7}{16}$	. 2734	48 350	176 900	6	......
172...	M-4..	.....do.........	.....do.........	6 by 19 by 6	$\frac{7}{16}$	. 2750	48 150	175 100	......	2
173...	M-4..	.....do.........	.....do.........	6 by 19 by 6	$\frac{7}{16}$	. 3001	56 000	186 700	1	......
174...	M-4..	.....do.........	.....do.........	6 by 19 by 6	$\frac{7}{16}$	. 2867	52 650	183 700	3	......
175...	M-4..	.....do.........	.....do.........	6 by 19 by 6	$\frac{7}{16}$	. 2912	52 010	178 600	2	......
176...	M-4..	.....do.........	.....do.........	6 by 19	$\frac{7}{16}$	. 3000	52 560	175 200	2	......
Mean						. 288	51 620	179 367		
177...	M-10.	Plow steel.....	Hoisting.......	6 by 19	½	. 3291	66 720	202 600	3	......
178...	M-10.	.....do.........	.....do.........	6 by 19	½	. 3291	66 880	203 200	3	......
179...	M-10.	.....do.........	.....do.........	6 by 19	½	. 3223	58 650	181 900	......	3
180...	M-10.	.....do.........	.....do.........	6 by 19	½	. 3223	56 570	175 500	......	1
181...	M-10.	.....do.........	.....do.........	6 by 19	½	. 3290	67 350	204 710	2	......
182...	M-10.	.....do.........	.....do.........	6 by 19	½	. 3290	71 550	217 480	3	......
183...	M-10.	.....do.........	.....do.........	6 by 19	½	. 3290	65 920	200 360	3	......
184...	M-10.	.....do.........	.....do.........	6 by 19	½	. 3290	66 250	201 370	4	......
185...	M-.9.	.....do.........	.....do.........	6 by 19	½	. 3081	72 300	234 700	......	4
Mean						. 325	65 799	202 424		
186...	M-4..	Plow steel.....	Hoisting.......	8 by 19	½	. 2563	52 560	205 100	3	......
187...	M-4..	.....do.........	.....do.........	8 by 19	½	. 2792	54 020	193 600	......	
188...	M-4..	.....do.........	.....do.........	8 by 19	½	. 2792	56 660	202 800	......	
189...	M-4..	.....do.........	.....do.........	8 by 19	½	. 2792	54 210	194 200	......	
190...	M-10.	.....do.........	.....do.........	8 by 19	½	. 3110	64 480	207 400	4	......
191...	M-10.	.....do.........	.....do.........	8 by 19	½	. 3110	67 620	217 500	4	......
192...	M-10.	.....do.........	.....do.........	8 by 19	½	. 3100	65 160	210 200	4	......
193...	M-10.	.....do.........	.....do.........	8 by 19	½	. 3100	61 380	198 000	......	1
194...	M-9..	.....do.........	.....do.........	8 by 19	½	. 2776	54 700	197 100	4	......
195...	M-9..	.....do.........	.....do.........	8 by 19	½	. 2647	55 550	209 900	......	4
Mean						. 288	58 634	203 580		
196...	M-10.	Plow steel.....	Steam shovel, etc.	6 by 37	½	. 3260	75 010	230 100	4	......

TABLE 9.—Tensile Strengths of 1-Inch Diameter Steel Cables

Serial No.	General data						Observed mechanical data			
	Manu-fac-turer	Type of steel	Use in practice	Strands and wires	Diam-eter of rope core	Sec-tional area ob-served	Maximum load		Number of strands broken	
									In body	At sock-et
					Inch	Inch ^{2}	Pounds	Lbs./in.^{2}		
197...	M-1..	Swedish iron..	Boat tillers....	6 by 42	$\frac{7}{8}$	0.2710	17 230	63 600	......	3
198...	M-1..	......do..........	......do..........	6 by 42	$\frac{7}{8}$	.2323	17 450	75 100	......	4
Mean						.252	17 340	69 350	......	......
199...	M-4..	Galvanized. steel.	Rigging guys, etc.	6 by 7	$\frac{7}{8}$	.4650	27 850	61 200	1	......
200...	M-4..	......do..........	......do..........	6 by 7	$\frac{7}{8}$	.3170	24 800	78 200	2	......
201...	M-4..	......do..........	......do..........	6 by 7	$\frac{7}{8}$	.3499	27 030	77 250	3	......
202...	M-10..	......do..........	......do..........	6 by 7	$\frac{7}{8}$	.3640	40 800	112 000	......	3½
Mean						.371	30 120	82 163	......	......
203...	M-4..	Crucible cast steel.	Hoisting	6 by 19 by 6	$\frac{7}{8}$	.3815	69 770	182 163	......	1
204...	M-4..	......do..........	......do..........	6 by 19 by 6	$\frac{7}{8}$	.3720	66 640	179 100	1	......
205...	M-4..	......do..........	......do..........	6 by 19 by 6	$\frac{7}{8}$	.3951	70 750	179 100	2	......
206...	M-2..	......do..........	......do..........	6 by 19	$\frac{7}{8}$	.3970	59 130	148 900	3	......
Mean						.386	66 573	172 500	......	......
207...	M-9..	Plow steel.....	Hoisting	6 by 19	$\frac{3}{8}$	.415	75 710	182 300	3	......
208...	M-9..	......do..........	......do..........	6 by 19	$\frac{3}{8}$	.415	76 270	183 800	2	......
Mean						.415	75 990	183 050	......	......
209...	M-4..	Plow steel.....	Extra flexible hoisting.	8 by 19	$\frac{7}{16}$	.3330	59 110	177 400	......	3
210...	M-4..	......do..........	......do..........	8 by 19	$\frac{9}{16}$	.3560	66 330	186 300	3	......
211...	M-4..	......do..........	......do..........	8 by 19 by 6	$\frac{7}{16}$	.3810	80 580	211 500	4	......
212...	M-4..	......do..........	......do..........	8 by 19	$\frac{7}{16}$	.3280	62 150	189 500	3	......
213...	M-11.	......do..........	......do..........	8 by 19	$\frac{7}{16}$	.3353	71 120	212 000	3	......
214...	M-11.	......do..........	......do..........	8 by 19	$\frac{7}{16}$	.3353	73 660	219 700	4	......
215...	M-10.	......do..........	......do..........	8 by 19	$\frac{7}{16}$	.3750	78 470	209 250	4	......
216...	M-10.	......do..........	......de..........	8 by 19	$\frac{7}{16}$	.3809	74 590	196 000	2	......
Mean						.353	70 751	200 206	......	......

Median curves (not shown in the figures) may be conceived to be drawn in such a way as to balance the "cuts" and "fills" of the serrated curves in the manner of fixing a railroad grade. These curves will correspond more approximately with least-square adjustments of the observations. It will be seen that there is a general tendency for the maximum stresses to occur at the smaller diameters of cables, with the exception of the guy ropes. After diameters of about three-fourths of an inch are reached the average stress, as determined by these median curves, becomes more nearly constant for any particular type of cable. The general form of the median curve is that shown for the upper curve drawn for the tiller rope, although the slopes are not as steep. This is also the form of the curves giving stresses in wires of different diameters.[6] The smallest wires of a particular steel develop the highest tensile stresses. This is commonly attributed to the effects of wiredrawing. It has been shown that the diameters of the wires in a cable are directly proportional to the diameters of the cables. It may, therefore, be inferred that the higher stresses developed in the smaller cables are due to the greater relative strengths of their constituent wires. Since relatively large wires are used in the guy ropes, evidence of this character is not presented.

TABLE 10.—Tensile Strengths of 1⅛-Inch Diameter Steel Cables

Serial No.	General data						Observed mechanical data			
	Manu-fac-turer	Type of steel	Use in practice	Strands and wires	Diam-eter of rope core	Sec-tional area ob-served	Maximum load		Number of strands broken	
									In body	At or near sock-et
					Inch	Inch ^{2}	Pounds	Lbs./in.^{2}		
217...	M-1..	Galvanized iron.	Rigging and guys.	6 by 7	⅝	0.5165	34 010	65 800	2	......
218...	M-1..	.....do.........	.....do.........	6 by 7	⅝	.4830	34 410	71 250	3	......
Mean	......	....	....	....	....	.500	34 210	68 525	......	......
219...	M-9..	Galvanized steel.	Rigging and guys.	6 by 7	½	.4990	52 350	104 900	3	......
220...	M-8..	.....do.........	.....do.........	6 by 7	½	.4596	36 650	79 700	1	......
Mean	......	....	....	....	....	.479	44 500	92 300	......	......
221...	M-4..	Crucible cast steel.	Hoisting	6 by 19 by 6	7/16	.4435	79 130	178 400	1	......
222...	M-4..	.....do.........	.....do.........	6 by 19 by 6	7/16	.4320	81 560	188 800	1	......
223...	M-4..	.....do.........	.....do.........	6 by 19 by 6	7/16	.4894	78 100	159 600	......	1
224...	M-4..	.....do.........	.....do.........	6 by 19 by 5	7/16	.4512	93 000	206 200	......	1
225...	M-4..	.....do.........	.....do.........	6 by 19 by 6	7/16	.4833	89 300	184 800	2	......
226...	M-2..	.....do.........	.....do.........	6 by 19 by 6	7/16	.4939	77 750	157 400	3	......
227...	M-4..	.....do.........	.....do.........	6 by 19 by 6	7/16	.4763	79 500	166 900	......	1
Mean	......	....	....	....	....	.467	82 620	177 443	......	......

[6] See report of Watertown Arsenal, Tests of Metals, p. 347; 1894; also Johnson's Materials of Construction, 4th ed., pp. 691–692; 1907.

TABLE 11.—Tensile Strengths of 1¼-Inch Diameter Steel Cables

Serial No.	Manufacturer	Type of steel	Use in practice	Strands and wires	Diameter of rope core	Sectional area observed	Maximum load		Number of strands broken	
									In body	At socket
					Inch	Inch²	Pounds	Lbs./in.²		
228...	M-4..	Crucible-cast steel.	Hoisting.......	6 by 19 by 6	⅝	0.6110	109 030	178 500	2	......
229...	M-9..	.....do........	.....do........	6 by 19	⅝	.6053	96 140	158 800	1	......
230...	M-4..	.....do........	.....do........	6 by 19 by 6	⅝	.6384	101 740	159 400	......	1
231...	M-4..	.....do........	.....do........	6 by 19 by 6	⅝	.6457	107 350	166 300	......	1
232...	M-4..	.....do........	.....do........	6 by 19 by 6	⅝	.6511	107 040	164 400	3	......
Mean						.630	104 260	165 480	......	
233...	M-2..	Plow steel.....	Hoisting.......	6 by 19	⁹⁄₁₆	.6020	108 000	179 700	......	
234...	M-3..	.....do........	.....do........	6 by 19	⁹⁄₁₆	.6270	123 800	197 400	3	......
235...	M-1..	.....do........	.....do........	6 by 19	⁹⁄₁₆	.6180	119 000	192 700	2	......
236...	M-6..	.....do........	.....do........	6 by 19	⁹⁄₁₆	.6350	130 000	204 800	......	2
237...	M-7..	.....do........	.....do........	6 by 19	⁹⁄₁₆	.6240	119 000	190 400	......	1
238...	M-4..	.....do........	.....do.......	6 by 19 by 6	⁹⁄₁₆	.6590	126 700	192 000	1	......
239...	M-5..	.....do........	.....do........	6 by 19	⁹⁄₁₆	.6220	125 500	201 500	1	......
240...	M-9..	.....do........	.....do........	6 by 19	⁹⁄₁₆	.6277	130 530	207 950	1	......
241...	M-10.	.....do........	.....do........	6 by 19	⁹⁄₁₆	.6650	140 700	211 580	......	1
242...	M-10.	.....do........	.....do........	6 by 19	⁹⁄₁₆	.6434	164 800	256 100	......	2
Mean						.632	128 803	203 413	......	
243...	M-10.	Plow steel.....	Extra flexible..	8 by 19	⅝	.5494	116 640	215 000	3	......

3. ANALYSIS OF FRACTURES

For the purpose of analysis, the fracture is stated to have occurred "at the socket" in Tables 3 to 13, where one or more strands break within 6 inches of the zinc; otherwise it is said to be "in the body" of the cable. For the cables of small diameters the number of breaks at the socket is somewhat greater than in the body. This is believed to be due to the annealing action on the small wires by the molten zinc used in socketing the ends of the cables. In the cables of larger diameters, on the other hand, the greater number of fractures are in the bodies of the specimens. It will be seen that in the case of breaks at the socket the maximum strengths compare very favorably with those in the body of the specimen. In those cases in which all the strands are broken, the fracture is usually near the socket, whereas in the body fractures two to four are usually broken. It may be inferred that a fracture at the socket should indicate slightly higher values in general than when the break occurs in the body of the specimen. The effects of heat should be considered in such cases.

TABLE 12.—Tensile Strengths of 1½-Inch Diameter Steel Cables

Serial No.	General data						Observed mechanical data			
	Manufacturer	Type of steel	Use in practice	Strands and wires	Diameter of rope core	Sectional area observed	Maximum load		Number of strands broken	
									In body	At socket
					Inch	Inch²	Pounds	Lbs./in.²		
244...	M-2..	Plow steel.....	Hoisting, etc...	6 by 19 by 6	11/16	0.9800	211 700	216 000	2	......
245...	M-4..	.....do........	.....do.........	6 by 19 by 6	11/16	.9430	183 180	194 300	1	......
246...	M-4..	.....do........	.....do.........	6 by 19 by 6	11/16	.9430	183 440	194 500	2	......
247...	M-4..	.....do........	.....do.........	6 by 19 by 6	11/16	.9430	189 900	201 400	......	1
248...	M-2..	.....do........	.....do.........	6 by 19	11/16	.8600	180 900	210 000	2	......
249...	M-9..	.....do........	.....do.........	6 by 19	11/16	.9030	176 500	195 100	......	3
250...	M-1..	.....do........	.....do.........	6 by 19	11/16	.9500	188 300	198 400	3	......
251...	M-8..	.....do........	.....do.........	6 by 19	11/16	.9060	148 000	163 500	2	......
252...	M-6..	.....do........	.....do.........	6 by 19	11/16	.9730	183 000	188 300	......	2
253...	M-7..	.....do........	.....do.........	6 by 19	11/16	.9730	184 500	189 500	......	2
254...	M-4..	.....do........	.....do.........	6 by 19	11/16	.9480	190 800	201 000	3	......
255...	M-5..	.....do........	.....do.........	6 by 19	11/16	.8880	163 500	184 500	2	......
256...	M-4..	.....do........	.....do.........	6 by 19 by 6	11/16	.9420	195 400	207 400	3	......
257...	M-4..	.....do........	.....do.........	6 by 19 by 6	11/16	.9460	192 300	203 300	2	......
258...	M-4..	.....do........	.....do.........	6 by 19 by 6	11/16	.9380	186 860	199 200	3	......
259...	M-2..	.....do........	.....do.........	[a]6 by 19 by 6	[a]¼	1.0620	200 800	189 100	3	......
260...	M-2..	.....do........	.....do.........	[a]6 by 19 by 6	[a]¼	1.0620	196 200	184 800	3	......
261...	M-4..	.....do........	.....do.........	6 by 19 by 6	11/16	.9380	187 000	199 400	3	......
262...	M-4..	.....do........	.....do.........	6 by 19 by 6	11/16	.9380	191 000	203 600	3	......
263...	M-2..	.....do........	.....do.........	[a]6 by 19 by 6	[a]¼	1.1370	230 260	202 500	3	......
264...	M-2..	.....do........	.....do.........	[a]6 by 19 by 6	[a]¼	1.1370	233 280	205 300	4	......
265...	M-2..	.....do........	.....do.........	6 by 19 by 6	11/16	1.0452	213 430	204 200	3	......
266...	M-1..	.....do........	.....do.........	6 by 19	[a]¼	1.1027	219 560	199 100	......	1
267...	M-2..	.....do........	.....do.........	[a]6 by 19	[a]¼	1.1430	230 480	201 600	3	......
268...	M-10.	.....do........	.....do.........	6 by 19	11/16	.9230	189 200	205 000	......	2
Mean	......	....	....	....	......	.981	193 940	197 640	......	......

[a] Special construction ¼-inch diameter rope core with 6 filler strands.

A comparison of the relative influences of the type of fracture in determining the ultimate strength of the specimen is found by regrouping the observed maximum stresses according to the two types of fractures and comparing. Taking, for example, the one-fourth-inch crucible cast steel cables, the values are 198 900 pounds per square inch for the body fractures and 198 500 pounds per square inch for the sockets, the mean falling between. For the seven-eighths-inch plow-steel hoisting cables the corresponding values are 204 953 and 197 367 per square inch, the mean for the set being 202 424, etc. If a law exists, the evidence would seem to indicate that the variation attributable to the local effects at the end connections are less than those which may be attributed to the differences in the particular grades of steel furnished by the manufacturers in meeting the provisions of the specifications.

TABLE 13.—Tensile Strengths of Cables of Diameters 2, 2¾, and 3¼ Inches of Plow Steel

| Serial no. | Diameter | Number of strands | | Wires in large strands | | Wires in small strands | | Sectional area | First report of wires breaking | Maximum load | | Number of strands broken | |
		Large	Small	Number	Diameter	Number	Diameter					In body	At socket
	Inches				Inch		Inch	Ins.²		Pounds	Lbs./in.²		
269	2	6	0	19	0.1285	0	....	1.55	280 000	293 000	189 000	1	......
				6	.0520								
270	2	6	0	19	.1360	0	....	1.65	255 000	304 000	184 200	......	1
271 b	2¾	6	0	19	.1890	0	....	3.36	540 000	550 000	163 400	1	......
				6	.0760								
272	2¾	6	0	19	.1850	0	....	3.23	645 000	694 000	214 900	2	......
				6	.0760								
273	3¼	6	6	25	.1540	9	0.0550	4.61	735 000	866 000	187 900	2	......
				6	.1695	10	.0935						
				6	.1285	....	....						
274	3¼	6	6	42	.1160	7	.0860	4.63	720 000	937 000	202 400	......	1
				6	.1360	6	.0820						
				6	.0995	6	.0670						
				7	.1285	....	....						

RESULTS OF CHEMICAL ANALYSES OF STEEL

| Serial No. | Diameter | Constituents | | | | |
		Si	P	C	S	Mn
	Inches	Per cent	Per cent	Per cent	Per cent	Per cent
269 c.........	2.	....	....	....	....	....
270.........	2	0.172	0.024	0.90	0.034	0.48
271.........	2¾	.130	.016	.68	.030	.58
272.........	2¾	.152	.027	.77	.036	.46
273.........	3¼	.160	.033	.58	.032	.41
274 d.........	3¼	.169	.019	.82	.025	.23

a For identification of manufacturer, see "Report of Tests of Engineering Record," July 15, 1916, p. 81. For "Use in practice," see abstract of specifications in Part I of this paper.
b Type of steel not stated.
c There was no analysis of this specimen.
d Qualitative tests for vanadium and nickel on Serial No. 274 gave negative results.

Most of the wires failed with the cup-shaped fractures common in ordinary steel specimens. The remainder failed with the angular fractures also found for steel bars. While a large number of the cables were examined for the particular type of wire fractures, in the opinion of the authors there was no law deducible for the type of fracture of the wires. The occurrence of angular fractures may, perhaps, be due to points of weakness in the "skin coat" or to planes of weakness of the material which develop under the influence of shearing forces coming into play as the helix adjusts itself in the strands.

4. ELONGATIONS AND REDUCTIONS IN DIAMETERS DISCUSSED

The percentages of elongation and the reductions in the diameters under cumulative loads for about 50 per cent of the test specimens were measured and are recorded in Tables 14 to 17. The reductions in the diameter as the loads increased were found with a Brown & Sharpe screw micrometer reading to 0.0001 of an inch. It was impracticable to measure elongations with an extensometer, as is done with a steel bar, on account of the twisting of the specimens, which increases with the load. This twisting occurs even with very short gage lengths and vitiates results. The elongations in these tests were measured directly with a graduated scale between two points on a cable, usually 80 cms apart. The elongations in inches were reduced to strains by dividing by the gage length. The results were multiplied by 100 to obtain the percentages of unit elongation.

Stress-strain data of the above description are seldom published. They are important in the development of the mechanics of the cable, the laws of which in the past have been rather difficult to formulate on account of the comparatively wide variations found in test results. The variableness in the elongations may be noted in the tables, even in those cables of the same size and construction.

In applying the methods of the theory of elasticity to the analyses of the behavior of the materials of construction, it is necessary to broadly differentiate between those whose structures are heterotropic (aeolotropic) and possess different elastic behaviors in various directions, as in the case of fibrous woods, and those which are isotropic, or in which the structure is elastically essentially homogeneous, as in steel. The structure of a cable considered as a unit is relatively complex when the arrangements and winding of its wires are considered. The analysis of the behavior becomes simplified, however, if the load is assumed to be uniformly distributed over the cross section for practical purposes, as is done in the case of a steel bar. It will be so treated in the present discussion. The question of the actual distributions of stresses in the wires is deferred to a later point in the analysis of test results.

TABLE 14.—Cumulative Elongations and Reductions in Diameters for Cables

¼-INCH CABLES a

Serial No.	Manufacturer	Type of steel	Strands and wires	Sectional area	Elongations in per cents at loads in pounds of—								Reduction in diameter in inches at loads in pounds of—							
					500	600	1000	1200	1500	1800	2000	3000	500	600	1000	1200	1500	1800	2000	3000
				Inch²																
1....	M-4..	Swedish iron.......	6 by 19	0.0174	0.130	......	0.500	......	0.750	......	0.940	......	0.000	......	0.002	......	0.004	......	0.005	......
4....	M-4..	.....do.....	6 by 42	.0160	.370	......	.870	......	1.500	......	2.000	......	.003	......	.011	......	.015	......	.019	......
5....	M-4..	.....do.....	6 by 42	.0179	......	0.620	......	1.230	......	1.910	......	......	......	0.005	......	0.009	......	0.012	......	......
6....	M-4..	.....do.....	6 by 42	.0160	......	.500	......	1.060	......	1.630	......	......	......	.004	......	.009	......	.014	......	......
Mean.....					......	.560	......	1.145	......	1.770	......	......	......	.005	......	.009	......	.013	......	......
8....	M-4..	Crucible cast steel...	6 by 19	.0232	.190	......	.380	......	.560	......	.750	1.190	.002	......	.004	......	.005	......	.006	0.008
9....	M-4..	.....do.....	6 by 19	.0228	......	.280	......	.560	......	.840	......	1.360	......	.003	......	.006	......	.009	......	.012
10...	M-4..	.....do.....	6 by 19	.0229	.060	......	.250	......	.500	......	.750	1.310	.002	......	.004	......	.005	......	.006	.008
11...	M-4..	.....do.....	6 by 19	.0229	.130	......	.310	......	.560	......	.810	1.190	.002	......	.005	......	.008	......	.010	.014
12...	M-4..	.....do.....	6 by 19	.0229	.130	......	.380	......	.560	......	.750	1.210	.005	......	.010	......	.011	......	.013	.015
13...	M-4..	.....do.....	6 by 19	.0259	.190	......	.340	......	.500	......	.620	.900	.001	......	.002	......	.003	......	.005	.007
14...	M-4..	.....do.....	6 by 19	.0259	.250	......	.400	......	.560	......	.690	.960	.001	......	.002	......	.003	......	.004	.005
15...	M-4..	.....do.....	6 by 19	.0259	.150	......	.340	......	.540	......	.710	1.060	.003	......	.004	......	.006	......	.008	.011
Mean.....					.157	......	.343	......	.540	......	.726	1.148	.002	......	.004	......	.006	......	.007	.010

¾-INCH CABLES ᵃ

Serial No.	Manu-facturer	Type of steel	Strands and wires	Sec-tional area	Elongations in per cents at loads in pounds of—								Reduction in diameter in inches at loads in pounds of—			
					1000	2000	3000	4000	5000	6000	7000	8000	1000	2000	3000	4000
				Inch²												
22	M-4	Swedish iron	6 by 19	0.0605	0.200	0.437	0.625	1.062					0.002	0.004	0.006	0.009
23	M-4	do	6 by 42	.0388	.350	.664	1.162	1.851					.005	.011	.017	.023
28	M-4	Galvanized steel	6 by 7	.0528	.125	.275	.625						.002	.004	.007	
29	M-4	do	6 by 7	.0552	.313	.500	1.000						.004	.005	.008	
30	M-4	do	6 by 7	.0552	.375	.625	1.125						.003	.008	.013	
31	M-4	do	6 by 7	.0526	.225	.400	.713						.001	.004	.008	
32	M-4	do	6 by 7	.0526	.375	.625	1.093						.004	.010	.015	
Mean					.283	.485	.911						.003	.006	.010	
34	M-4	Crucible cast steel	6 by 19	.0563	.250	.475	.650	.837	1.000	1.190	1.375	1.562				
35	M-4	do	6 by 19	.0567			.787	1.020							.015	.019

ᵃ An initial load of 100 pounds was taken as the zero of observations for these cables.

TABLE 14.—Cumulative Elongations and Reductions in Diameters for Cables—Continued

½-INCH CABLES a

Serial No.	Description of cable				Elongations in per cents at loads in pounds of—								Reduction in diameter in inches at loads in pounds of—							
	Manufacturer	Type of steel	Strands and wires	Sectional area	1000	2000	3000	4000	5000	6000	10 000	14 000	1000	2000	3000	4000	5000	6000	10 000	14 000
				Inch²																
46	M-4	Swedish iron	6 by 42	0.0781	0.251		1.002			2.500			0.002		0.002			0.035		
47	M-4	do	6 by 42	.0640			1.000			2.750					.018			.034		
48	M-4	do	6 by 42	.0714			.887			1.625				.011	.014			.026		
49	M-4	do	6 by 42	.0597			.750			1.450					.011			.019		
50	M-4	do	6 by 42	.0641			.987			2.125					.019			.033		
Mean							.925			2.090					.013			.029		
51	M-4	Galvanized steel	6 by 7	.1010	.113	0.238	.375	0.513	0.675	.863			.001	.003	.005	0.007	0.009	.012		
52	M-4	do	6 by 7	.0892		.338	.450		.862	.938				.011	.011		.016	.018		
53	M-4	do	6 by 7	.0892	.162		.525	.687		1.625			.003	.008	.012	.014		.023		
Mean					.138	.288	.450	.600	.769	1.142			.002	.006	.009	.011	.013	.018		
55	M-4	Crucible cast steel	6 by 9	.1050			.326			.688	1.112	1.700			.006			.011	0.015	0.020
56	M-4	do	6 by 9	.1035			.413			.750	1.270	1.625			.007			.014	.021	.024
57	M-4	do	6 by 9	.0971	.125	.262	.437	.687	.875	1.060	1.560	2.070	.002	.003	.005	.008	.010	.013	.020	.024
Mean							.392			.833	1.314	1.798			.006			.013	.019	.022

a The initial load was 200 pounds for these cables.

TABLE 15.—Cumulative Elongations and Reductions in Diameters for ⅝-Inch Cables[a]

Serial No.	Manufacturer	Type of steel	Strands and wires	Sectional area	Elongations in per cents at loads in pounds of—								Reduction in diameter in inches at loads in pounds of—							
				Inch²	2000	4000	6000	8000	10000	12000	18000	24000	2000	4000	6000	8000	10000	12000	18000	24000
66	M-4	Galvanized	6 by 7	0.1664	0.125	0.225	0.288	0.425	0.650	2.250			0.004	0.007	0.009	0.011		.011	.015	
67	M-4	do.	6 by 7	.1570	.250	.375	.500	.625	.875	2.750			.001	.004	.008	.010	0.012	.013	.018	
68	M-4	do.	6 by 7	.1616	.212	.400	.538	.650	1.000				.005	.009	.012	.014	.021			.023
Mean					.196	.333	.442	.567	.842	2.500			.003	.007	.010	.012	.017	.012	.017	.023
70	M-4	Crucible cast	6 by 19	.1466	.125	.250	.375	.500	.625	.750	1.125		.001	.003	.005	.007	.009	.011	.015	
71	M-4	do.	6 by 19	.1550	.250	.375	.500	.625	.812	.875	1.350		.003	.005	.007	.009	.011	.013	.018	
72	M-4	do.	6 by 19	.1570	.187	.313	.438	.562	.688	.812	1.250		.002	.005	.007	.008	.009	.011	.015	
73	M-4	do.	6 by 19 by 6	.1648			.500			.825	1.187	2.250			.005			.017	.022	.023
74	M-4	do.	6 by 19 by 6	.1648			.375			.750	1.250	2.550			.009			.014	.018	.026
75	M-4	do.	6 by 19	.1626			.375			.750	1.125	1.750			.014			.020	.025	.030
76	M-4	do.	6 by 19 by 6	.1651		.312		.500		.700		1.688				.009		.012		.020
77	M-4	do.	6 by 19 by 6	.1651		.313		.587		.875		1.813				.017		.020		.032
Mean					.187	.313	.427	.555	.708	.792	1.215	2.010	.002	.004	.008	.010	.010	.015	.019	.026
101	M-9	Plow	6 by 19	.1590	.288	.501	.688	.788	.938	1.062	1.438		.005	.009	.011	.013	.015	.017	.023	
102	M-9	do.	6 by 19	.1510	.213	.363	.525	.675	.813	.937	1.375	1.788	.005	.007	.009	.012	.014	.018	.024	.031
103	M-9	do.	6 by 19	.1510	.150	.275	.388	.525	.650	.775	1.125	1.488	.005	.008	.011	.014	.016	.018	.024	.030
104	M-2	do.	6 by 19	.1432			.400			.825	1.250	1.876			.006			.011	.015	.020
Mean					.217	.380	.500	.663	.800	.900	1.297	1.717	.005	.008	.009	.013	.015	.016	.022	.027
78	M-1	Plow	6 by 19	.1505	.235	.469	.678	.814	.975	1.135	1.568	2.037	.005	.010	.016	.021	.025	.026	.032	.036
79	M-1	do.	6 by 19	.1433			.587			.875	1.187	1.588			.014			.020	.023	.026
80	M-1	do.	6 by 19	.1433		.437	.587	.737	.875	1.000	1.300	1.725		.009	.014	.018	.024	.027	.035	.040
81	M-2	do.	6 by 19 by 6	.1444			.750			1.250	1.788	2.376			.014			.020	.025	.030
82	M-2	do.	6 by 19	.1469			.687			1.312	1.875	2.435			.017			.027	.032	.038
83	M-4	do.	6 by 19 by 6	.1677			.437			.812	1.125	1.749			.010			.017	.021	.024

[a] An initial load of 500 pounds was taken as the zero of observations for these cables.

Table 15.—Cumulative Elongations and Reductions in Diameters for ⅝-Inch Cables[a]—Continued

Serial No.	Manufacturer	Description of cable			Elongation in per cents at loads in pounds of—								Reduction in diameter in inches at loads in pounds of—							
		Type of steel	Strands and wires	Sectional area	2000	4000	6000	8000	10000	12000	18000	24000	2000	4000	6000	8000	10000	12000	18000	24000
84	M-4	Plow	6 by 19 by 6	0.1677			0.593			1.050	1.360	1.730			0.016			0.027	0.030	0.035
85	M-9	do.	6 by 19	.1521			.463			.813	1.188	1.752			.009			.014	.019	.025
86	M-8	do.	6 by 19	.1621			.432			.679	1.012	1.335			.005			.009	.012	.014
87	M-2	do.	6 by 19	.1579			.338			.675	1.063	1.576			.004			.007	.010	.014
88	M-1	do.	6 by 19	.1433			.438			.775	1.168	1.500			.010			.015	.025	.029
89	M-2	do.	6 by 19	.1579			.313			.650	1.000	1.500			.003			.010	.015	.021
90	M-2	do.	6 by 19	.1579			.313			.625	1.000	1.628			.004			.010	.015	.021
91	M-2	do.	6 by 19	.1579			.407			.741	1.285	1.975			.004			.010	.015	.023
92	M-2	do.	6 by 19	.1579			.275			.749	1.185	1.870			.004			.008	.012	.018
93	M-2	do.	6 by 19	.1579			.312			.749	1.125	1.623			.008			.014	.020	.027
94	M-9	do.	6 by 19	.1579			.500			1.000	1.420	2.060			.004			.008	.013	.018
95	M-5	do.	6 by 19	.1485			.625			1.000	1.375	1.938			.009			.014	.018	.021
96	M-5	do.	6 by 19	.1485			.574			.874	1.186	1.622			.010			.013	.016	.020
Mean						0.453	.491	.775	.925	.882	1.274	1.791		0.010	.009	.020	.025	.016	.020	.027
105	M-4	Plow, extra flexible	8 by 19	.1280	0.187	.338	.537	.812	1.062	1.225	1.750		0.002	.006	.010	.014	.017	.019	.025	
106	M-4	do.	8 by 19	.1070	.263	.563	.938	1.250	1.390	1.538	2.125		.004	.017	.028	.034	.036	.040	.050	
107	M-9	do.	8 by 19	.1398			.500			.850	1.175				.007			.014	.018	
108	M-4	do.	8 by 19	.1502			.568			1.000	1.438	1.999			.010			.021	.029	.035
109	M-4	do.	8 by 19	.1381			.725			1.250	1.750	2.376			.016			.022	.027	.031
110	M-4	do.	8 by 19	.1460			.625			1.125	1.625	2.313			.007			.014	.021	.028
111	M-4	do.	8 by 19	.1460			.494			.865	1.235	1.852			.011			.018	.024	.030
112	M-4	do.	8 by 19	.1240		.400		.900		1.212				.004		.027		.034		
113	M-4	do.	8 by 19	.1208			.875			1.625	2.125	3.060			.033			.047		
114	M-4	do.	8 by 19 by 0	.1374			1.000			1.525	2.035	2.748			.019			.030	.036	.041
Mean					0.225	.434	.696	.987	1.226	1.222	1.695	2.391	0.003	.009	.016	.025	.027	.026	.029	.033

[a] An initial load of 500 pounds was taken as the zero of observations for these cables.

TABLE 16.—Cumulative Elongations and Reductions in Diameters for Cables

¾-INCH CABLES[a]

Serial No.	Manufacturer	Type of steel	Strands and wires	Sectional area (Inch²)	Elongations in per cents at loads in pounds of—								Reduction in diameter in inches at loads in pounds of—							
					3000	6000	9000	12000	15000	18000	24000	30000	3000	6000	9000	12000	15000	18000	24000	30000
122	M-4..	Galvanized	6 by 7	0.1905	0.250	0.437	0.562	1.810					0.008	0.014	0.020	0.035				
125	M-4..	Crucible cast.	6 by 19	.2173	.188	.300	.375	.462	0.562	0.687	0.938	1.540	.008	.011	.012	.014	0.016	0.018	0.023	
126	M-4..	do	6 by 19	.2130	.250	.400	.587	.774	.937	1.087	1.400	1.870	.003	.009	.015	.022	.026	.029	.033	.037
127	M-4..	do	6 by 19	.2130	.250	.437	.562	.738	.875	1.000	1.275	1.775	.002	.007	.012	.015	.017	.020	.023	.027
128	M-4..	do	6 by 19 by 6	.2486	.150	.375	.600	.750	.875	1.000	1.250	1.550	.007	.015	.023	.030	.034	.036	.041	.045
129	M-4..	do	6 by 19 by 6	.2400		.333		.580		.802	1.035	1.555		.006		.012		.016	.020	.024
130	M-4..	do	6 by 19 by 6	.2400	.250	.377		.587		.812	1.173	1.560		.007		.011		.015	.019	.023
131	M-j..	do	6 by 19 by 6	.2405		.374	.500	.623	.787	.936	1.185	1.560	.005	.012	.017	.020	.023	.025	.028	.023
132	M-4..	do	6 by 19 by 6	.2376				.654			1.132	1.560				.028			.035	.033
133	M-4..	do	6 by 19 by 6	.2351				.500			1.250					.021			.032	
134	M-4..	do	6 by 19 by 6	.2351				.750			1.312					.026			.038	
Mean					.218	.371	.525	.642	.807	.903	1.195	1.630	.005	.010	.016	.020	.023	.023	.030	.032
135	M-4..	Plow	6 by 19	.2350	.125	.312	.475	.612	.787	.862	1.100	1.350	.005	.013	.014	.016	.018	.020	.023	.026
136	M-4..	do	6 by 19	.2180	.187	.375	.437	.687	.812	.950	1.290	1.660	.003	.007	.010	.013	.015	.017	.019	.021
137	M-4..	do	6 by 19	.2155	.250	.412	.574	.724	.850	.973	1.225	1.575	.006	.008	.011	.013	.016	.019	.024	.028
138	M-2..	do	6 by 19 by 6	.2176				.567			1.197					.012			.018	
139	M-1..	do	6 by 19 by 6	.2063	.123	.308	.432	.518	.603	.702	.950	1.230	.003	.006	.009	.010	.012	.013	.015	.017
140	M-1..	do	6 by 19	.2329				.382			.802					.010			.014	
141	M-4..	do	6 by 19 by 6	.2387				.556			1.048					.011			.017	
142	M-9..	do	6 by 19	.2174				.618			1.172					.012			.019	
143	M-8..	do	6 by 19	.2315				.617			1.170					.011			.017	
Mean					.171	.352	.480	.587	.763	.872	1.106	1.454	.004	.009	.011	.012	.015	.018	.018	.023

[a] An initial load of 500 pounds was taken as the zero of observations for these cables.

TABLE 16.—Cumulative Elongations and Reductions in Diameters for Cables—Continued

¾-INCH CABLES[a]—Continued

Serial No.	Manufacturer	Type of steel	Strands and wires	Sectional area (Inch²)	Elong. 3000	6000	9000	12000	15000	18000	24000	30000	Red. 3000	6000	9000	12000	15000	18000	24000	30000
149	M-4	Plow	8 by 9	0.1780	0.375	0.812	1.087	1.437	1.625	1.822	2.222	2.810	0.006	0.017	0.028	0.033	0.037	0.039	0.043	0.048
150	M-9	do	8 by 19	.2065		.400				1.000	1.200	1.500		.007				.014	.017	.021
151	M-4	do	8 by 19	.2047		.678		1.295		1.665	2.100	2.680		.009		.017		.023	.027	.032
152	M-4	do	8 by 19	.2047		.517		1.110		1.523		2.470		.009		.015		.021	.025	.030
153	M-4	do	8 by 19 by 6	.2224				1.062			1.725					.019			.031	
154	M-4	do	8 by 19 by 6	.2224				.937			1.500					.013			.022	
155	M-4	do	8 by 19 by 6	.1988				.812			1.312					.017			.026	
156	M-4	do	8 by 19 by 6	.2215				.875			1.475					.014			.023	
Mean						.601		1.075		1.503	1.687	2,365		.011		.018		.024	.027	.033

⅞-INCH CABLES[a]

Serial No.	Manufacturer	Type of steel	Strands and wires	Sectional area (Inch²)	Elong. 5000	10000	15000	20000	25000	30000	35000	40000	Red. 5000	10000	15000	20000	25000	30000	35000	40000
168	M-4	Galvanized	6 by 7	0.2640	0.250	0.437	0.623	3.310					0.010	0.015	0.024					
171	M-4	Crucible cast	6 by 19 by 6	.2734	.187	.374	.587	.736	0.936	1.073	1.387	1.685	.010	.019	.027	.034	.038	.041	.044	.046
172	M-4	do	6 by 19 by 6	.2750	.362	.538	.688	.874	1.063	1.250	1.448	1.773	.006	.014	.020	.026	.030	.034	.039	.044
173	M-4	do	6 by 19 by 6	.3001	.374	.686	.872	1.060	1.245	1.410	1.570	1.810	.011	.026	.034	.041	.043	.047	.051	.054
174	M-4	do	6 by 19 by 6	.2867		.474	.740	.812		1.123		1.623		.023	.029	.033		.039		.044
175	M-4	do	6 by 19 by 6	.2912	.247	.494	.750	.926	1.048	1.235	1.420	1.668	.011	.022	.029	.033	.036	.039	.041	.044
176	M-4	do	6 by 19	.3000	.275	.523		.973	1.100	1.248	1.440	1.688	.015	.031		.049	.053	.057	.061	.065
Mean					.289	.515	.727	.897	1.078	1.223	1.453	1.708	.011	.023	.031	.036	.040	.043	.045	.050
186	M-4	Plow	8 by 19	.2563		.862	.450	1.358		1.680		2.220		.028		.037		.045		.049
196	M-4	do	6 by 37	.3260	.150	.325		.600	.737	.862		1.110	.004	.006		.011	.012	.013		.017

a An initial load of 500 pounds was taken as the zero of observation for these cables.

TABLE 17.—Cumulative Elongations and Reductions in Diameters for Cables [a]

1-INCH CABLES

| Description of cable | | | | | Elongations in per cents at loads in pounds of [a] | | | | | | | | Reduction in diameter in inches at loads in pounds of [a] | | | | | | | |
Serial No.	Manufacturer	Type of steel	Strands and wires	Sectional area	3000	6000	9000	12000	15000	30000	36000	45000	3000	6000	9000	12000	15000	30000	36000	45000
				Inch²																
197	M-1	Swedish iron	6 by 42	0.2710	0.374	0.712	1.023	1.438	2.250	.....	.....	.....	0.016	0.028	0.034	0.043	0.061	.....	.....	.....
198	M-1	do	6 by 42	.2323	.....	.936	.....	1.935	3.460	.....	.....	.....	.....	.025	.....	.047	.073	.....	.....	.....
Mean					.374	.824	1.023	1.687	2.855	.....	.....	.....	.016	.027	.034	.045	.067	.....	.....	.....
199	M-4	Galvanized	6 by 7	.4550	.125	.287	.374	.500	.587	.....	.....	.....	.009	.020	.030	.037	.038	.....	.....	.....
200	M-4	do	6 by 7	.3170	.250	.500	.687	.812	.937	.....	.....	.....	.014	.028	.036	.042	.046	.....	.....	.....
Mean					.188	.394	.531	.656	.762	.....	.....	.....	.012	.024	.033	.040	.042	.....	.....	.....
203	M-4	Crucible cast	6 by 19 by 6	.3815	.125	.250	.375	.500	.600	0.936	1.063		.003	.008	.013	.020	.025	0.034	0.037	
204	M-4	do	6 by 19 by 6	.3720	.125	.250	.312	.400	.500	.925	1.087	1.325	.003	.005	.009	.012	.016	.028	.032	0.036
205	M-4	do	6 by 19 by 6	.3951	.....	.250	.....	.432	.....	.802	.....	1.172	.....	.....	.012	.021	.....	.033	.....	.042
Mean					.125	.250	.344	.444	.550	.888	1.075	1.249	.003	.007	.011	.018	.021	.032	.035	.039
209	M-4	Plow	8 by 19	.3330	.125	.250	.400	.523	.662	1.173	1.360	1.683	.003	.006	.010	.013	.015	.024	.028	.033
210	M-4	do	8 by 19	.3560	.375	.250	.649	.....	.873	.....	.....	1.848	.007	.....	.012	.....	.014	.....	.....	.026
212	M-4	do	8 by 19	.3280	.....	.....	.....	.....	.937	1.400	.....	2.087	.....	.....	.....	.....	.033	.044	.....	.....
Mean					.250	.250	.525	.523	.824	1.287	1.360	1.873	.005	.006	.011	.013	.021	.034	.028	.030

[a] An initial load of 500 pounds was taken as the zero of observations for 1-inch cables.

TABLE 17.—Cumulative Elongations and Reductions in Diameters for Cables—Continued

1⅛-INCH CABLES

Serial No.	Manufacturer	Type of steel	Strands and wires	Sectional area (Inch²)	Elongations in per cents at loads in pounds of a—								Reduction in diameter in inches at loads in pounds of a—							
					5000	10 000	15 000	20 000	30 000	40 000	50 000	60 000	5000	10 000	15 000	20 000	30 000	40 000	50 000	60 000
221	M-4.	Crucible cast	6 by 19 by 6	0.4435	0.125	0.275	0.450	0.624	0.874	1.125	1.375	1.687	0.005	0.013	0.022	0.028	0.038	0.046	0.051	0.056
222	M-4.	do	6 by 19 by 6	.4320	.125	.250	.375	.500	.712	.900	1.100	1.400	.004	.006	.008	.011	.016	.021	.026	.031
223	M-4.	do	6 by 19 by 6	.4894	.125	.250	.350	.450	.649	.737	1.062	1.312	.005	.010	.013	.018	.025	.029	.033	.....
224	M-4.	do	6 by 19 by 6	.4512	.125	.262	.400	.523	.750	.938	1.075	1.248	.005	.014	.021	.027	.036	.042	.047	.052
Mean					.125	.259	.394	.524	.746	.925	1.153	1.412	.005	.011	.016	.021	.029	.035	.039	.046

1¼-INCH CABLES

Serial No.	Manufacturer	Type of steel	Strands and wires	Sectional area (Inch²)	Elongations in per cents at loads in pounds of a—								Reduction in diameter in inches at loads in pounds of a—							
					10 000	20 000	30 000	40 000	50 000	60 000	80 000	100 000	10 000	20 000	30 000	40 000	50 000	60 000	80 000	100 000
228	M-4.	Crucible cast	6 by 19 by 6	0.6110	0.187	0.324	0.512	0.687	0.849	1.000	1.373		0.008	0.015	0.020	0.025	0.029	0.033	0.041	
229	M-9.	do	6 by 19	.6053	.300	.624	.750	1.063	1.237	1.475	2.187		.013	.025	.032	.041	.046	.051	.064	
230	M-4.	do	6 by 19 by 6	.6384	.148	.333	.562	.617	.740	1.062			.009	.018	.032	.028	.031	.045	.038	
231	M-4.	do	6 by 19 by 6	.6457	.212	.427	.608	.789	.942	1.116	1.357	1.639	.010	.019	.027	.031	.035	.041	.046	
Mean																				
233	M-2.	Plow	6 by 19	.6020	.270	.530	.800	1.000	1.200	1.400	2.000		.008	.013	.020	.028	.033	.036	.045	
234	M-3.	do	6 by 19	.6270	.200	.470	.670	.800	1.000	1.130	1.730	2.330	.006	.010	.016	.019	.025	.030	.041	.054
235	M-1.	do	6 by 19	.6180	.270	.400	.530	.670	.870	1.070	1.470	1.800	.006	.012	.019	.023	.025	.029	.035	.047

Serial No.	Manuf.	Type of steel	Strands and wires	Sectional area	Elong. 10 000	20 000	30 000	40 000	60 000	80 000	100 000	150 000	Red. 10 000	20 000	30 000	40 000	60 000	80 000	100 000	150 000
236	M-6	do	6 by 19	.6350	.270	.530	.800	1.070	1.200	1.200	1.600	2.000	.012	.025	.036	.043	.047	.050	.057	.062
237	M-7	do	6 by 19	.6240	.330	.670	1.000	1.270	1.600	2.000	2.400	2.930	.031	.031	.037	.047	.051	.054	.057	.063
238	M-4	do	6 by 19 by 6	.6590	.200	.400	.530	.670	.800	1.000	1.330	1.800	.003	.007	.011	.014	.017	.019	.027	.035
239	M-5	do	6 by 19	.6220	.270	.600	.730	1.000	1.130	1.330	1.730	2.200	.015	.035	.046	.059	.067	.073	.085	.097
Mean					.259	.514	.723	.926	1.114	1.304	1.751	2.177	.012	.019	.026	.033	.038	.042	.050	.060

1½-INCH CABLES

Serial No.	Manufacturer	Description of cable — Type of steel	Strands and wires	Sectional area (Inch²)	Elongations in per cents at loads in pounds of[a] — 10 000	20 000	30 000	40 000	60 000	80 000	100 000	150 000	Reduction in diameter in inches at loads in pounds of[a] — 10 000	20 000	30 000	40 000	60 000	80 000	100 000	150 000
244	M-2	Plow	6 by 19 by 6	0.9800	0.125	0.312	0.436	0.562	0.748	0.872	0.998	1.333	0.005	0.011	0.018	0.024	0.028	0.033	0.038	0.48
245	M-4	do	6 by 19 by 6	.9430	.062	.137	.237	.361	.598	.772	.972	1.632	.006	.011	.016	.019	.024	.029	.033	.045
246	M-4	do	6 by 19 by 6	.9430	.187	.362	.437	.562	.750	.912	1.125	1.873	.006	.012	.015	.018	.024	.029	.033	.045
248	M-2	do	6 by 19	.8600	.130	.270	.470	.600	.870	1.130	1.400	2.330	.004	.010	.017	.023	.032	.042	.052	.076
249	M-9	do	6 by 19	.9030	.130	.270	.400	.530	.800	1.130	1.400	2.270	.004	.012	.019	.025	.036	.045	.054	.073
250	M-1	do	6 by 19	.9500	.130	.270	.400	.530	.800	1.000	1.200	2.070	.007	.014	.024	.034	.047	.053	.060	.076
251	M-8	do	6 by 19	.9060	.130	.270	.400	.530	.800	1.070	1.333		.003	.007	.010	.012	.018	.020	.025	
252	M-6	do	6 by 19	.9730	.200	.330	.470	.600	.870	1.130	1.400	2.270	.004	.008	.012	.014	.020	.025	.031	.045
253	M-7	do	6 by 19	.9730	.130	.270	.400	.530	.800	1.000	1.270	2.130	.004	.011	.016	.021	.027	.034	.038	.053
255	M-5	do	6 by 19	.8880	.130	.330	.470	.670	.930	1.130	1.400	2.930	.007	.015	.023	.025	.031	.036	.042	.065
Mean					.135	.282	.412	.548	.804	1.015	1.250	1.884	.005	.011	.017	.022	.029	.035	.040	.053

[a] The initial load was 1000 pounds for 1⅛ inch and 1¼ inch cables, and 2000 pounds for 1½ inch cables.

In the preliminary discussion with manufacturers and engineers relative to the formulation of the program of this research, it was requested that careful consideration be given to the determina-

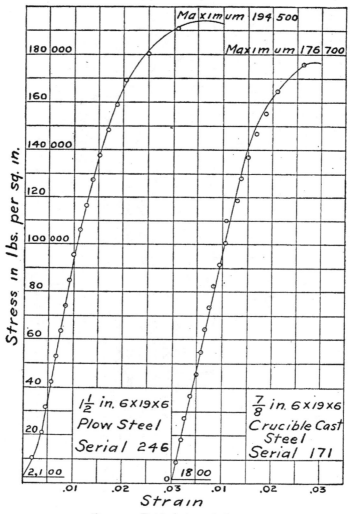

FIG. 11.—*Typical stress-strain curves*

The curve at the left, with the inflection, is more characteristic of the larger, stiffer specimens; that at the right, of the smaller, flexible specimens

tion of the modulus of elasticity of wire rope. The additional stress developed in the wires when a rope is bent over a sheave depends on this factor. The elongation data will be discussed from this point of view.

5. STRESS-STRAIN CURVES AND MODULI OF CABLES

An examination of the data of Tables 14 to 17 show that while the elongation at a certain load is somewhat variable for different test specimens of the same class, it is closely proportional to the load in any particular cable for a range up to about two-thirds of the ultimate strength. The typical forms of the stress-strain curves are shown in Fig. 11. The general type of curve is given in the case of the 1½-inch plow-steel cable of serial No. 246, where the form on account of the lateral stiffness of the specimen is somewhat exaggerated. Initial loads were taken, depending on the diameter of the cables, as are noted in the tables, in order to bring the specimens to firm bearings in the testing machine before measurements were made.

In the vicinity of the origin the curves are in general slightly convex to the horizontal axis. This portion of the curve is characteristic of the curves for matters whose densities increase with the applied stresses, as in the case of cellular and granular materials. In the present tests it is indicative of the fact that there are initial curvatures in the strands and wires from the laying, these not being in the most compact position, and probably a certain degree of "slack" or curvature in the cable itself. The elongation under stress for the lower loads, in other words, is not wholly elastic in the commonly accepted sense. As the load increases the strands and wires bed more firmly upon each other and upon the rope core, so that the parts are in more intimate contact. There then occurs a period of elastic behavior, in which the lower limit of proportionality is not very clearly defined on account of the attendant difficulties in making the measurements. The upper proportional limit, also not well defined, will correspond with the proportional limit of the steel in the wires, but it will be at a lower relative height on the curves, as a rule, than for the wires, as there are probably slight slippings and readjustments under stress still occurring among the strands and wires. The behavior of the cable, in other words, on account of its structure is not quite so uniform as that of a homogeneous bar, where the structure of the metal may be assumed to be continuous.

The curve for the seven-eighths inch crucible cast steel cable, serial No. 171, shown in Fig. 11, is more nearly straight, since the cable is more flexible and the lateral stiffness of the wires is less, on account of their smaller diameters. The contraflexure, as a rule, is not so pronounced in the smaller diameters of cables, although at times it is still in evidence. The curve shown is more characteristic of the curves for wires.

When the load is reduced from the upper proportional limit back to the initial point, the strain energy, which is absorbed in stretching, twisting, and bending the wires and strands, considering them as spiral springs, is partially released. A portion of the total energy has been transformed in permanently compressing the rope core and in bringing the wires into a more dense configuration. The result on reversal of the load is apparent as a set. A succession of applications and reversals of the loads shows the law of diminishing sets, the successive stress-strain curves increasing somewhat in slope. It may be inferred, generally speaking, that the elasticity of the cable is improved by use when the working stresses are not exceeded, as is often the case with homogeneous materials of relatively continuous structure—for example, steel and cast iron.

When a piece of a cable is clamped in a vise and a transverse force is applied at one end, it behaves as a straight flat spring. When, however, the force is removed, the specimen does not return quite to the unstrained position. With the application of a slight upward force it resumes and remains at this position. The operations may be repeated successively with like results. This shows that the interstitial frictions and the interactions of the constituent wires are also in part responsible for the sets. The longitudinal shearing stresses in a solid bar of steel are not thus overcome, and the modulus is correspondingly much higher.

It was found necessary to make an independent investigation of Unwin's modulus, which is employed when sets are considered, on account of the complex character of the phenomena that are presented and the number of reversals of loading which must be used in investigating the problem by purely static methods of loading. Actual service tests of cables over sheaves are most appropriate for studying Unwin's modulus. Young's modulus will be employed in the following discussion of bending.

6. THE LAW OF BENDING

The tables of bending stresses calculated by wire-rope manufacturers when the ropes are bent over sheaves of different diameters are determined upon the assumption of elastic behavior of the cables.[7] The relation between the bending couple and the curva-

---

[7] American Wire Rope; 1913. Handbook issued by the American Steel & Wire Co., p. 32 et seq.

ture as given in works on mechanics is $\frac{M}{EI}=\frac{1}{r}$: M is the bending couple, E is Young's modulus of elasticity, I is the moment of inertia for the cross section of the cable, and r is the radius of curvature of the bend. Let B be the diameter of the bend, D the diameter of the cable, and f the bending stress at the outer wires, which are most stressed. The stress in the outer wire varies as in the law of the beam, viz, $f=\frac{M}{I}\frac{D}{2}$. There results from these equations the expression for bending stress in the cable, $f=E\frac{D}{B}$; i. e., the stress is proportional to the modulus and the diameter of the cable. It is inversely proportional to the diameter of the sheave. Accordingly, the manufacturers recommend as large a sheave as is consistent with mechanical design. They seek to determine the upper limit of the modulus of the cable.

7. YOUNG'S MODULUS OF CABLES

The modulus is the ratio of the stress to the strain, taken within the limit of proportionality. The loads given in Tables 14 to 17 were divided by the areas of the cables as given in the table for obtaining the stresses. The unit elongations percentages as given in the table of elongations were divided by 100 to give the strains, and the ratio of the stress to strain calculated. Three to five determinations were made in this manner for each cable considered. The total averages were found by weighting the averages for each particular cable by the number of separate calculations. The results are plotted in Fig. 12.

It is seen from the figure that in general the modulus is somewhat higher for small diameters of cables than it is for the larger diameters. This may be influenced by the fact that the stresses in small wires are higher, as has been remarked. It is probably mainly due to the fact that for the small diameters of cables the influence of slack and initial curvatures in the cable strands and wires is not so pronounced. The cables being more flexible they behave more as cords. Free elastic action is less restricted. The lengthening due to the "straightening out" of curvature is not as much in evidence as for the stiffer specimens.

A number of tables of bending stresses have been calculated upon the assumption that the modulus of new wire rope does not exceed 12 000 000 pounds per square inch of cable section.[8] The

---

[8] See American Wire Rope, p. 33; 1913.

values of Fig. 12 indicate that this is a conservative figure. The values deduced under a series of repeated load and impact tests will probably increase the values in the figure to some extent. For working stresses of one-seventh to one-fifth the breaking load, it is believed, the modulus of 12 000 000 pounds per square inch will insure a sufficient safe margin in the calculation of bending

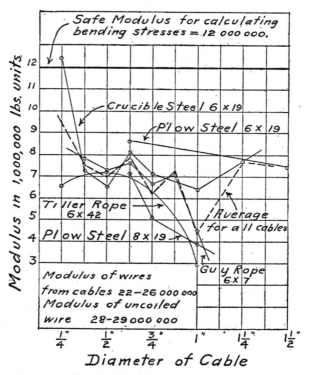

Fig. 12.—*Moduli of elasticity for cables of different types and diameters*

stresses until more detailed information can be furnished from tests made under kinetic conditions of loading, and a study of the behaviors of bending over sheaves in service tests is possible.

VII. QUALITIES OF MATERIAL IN PLOW-STEEL CABLES

In view of the relative importance of plow steel for hoisting rope, and the fact that the larger number of cables tested were of this steel, a comparative analysis of the materials of plow-steel ropes furnished by different manufacturers was made as the basis of the investigation. The data given upon fibers and lubricants of rope cores are doubtless fairly representative of the entire series of cables. While the chemical constituents of the steel will vary in different cables, the range of departures from the mean of the constituents is probably quite typical of that which is generally found for cables of the same class from different manufacturers and similarly for the physical properties and coefficients.

TABLE 18.—Results of Chemical Analyses of Steel in 6 by 19 Plow-Steel Cables

1¼-INCH PLOW-STEEL CABLES

Serial No.	Manu-fac-turer	Position of wire in strand a	Si	P	C	S	Mn
233	M-2...	1.................	0.12	0.053	0.64	0.058	0.46
		2 and 3.................	.12	.052	.80	.029	.55
234	M-3...	1, 2, and 3.................	.015	.044	.66	.078	.37
235	M-1...	1.................	.03	.044	.13	.032	.32
		2 and 3.................	.16	.040	.82	.037	.31
236	M-6...	1.................	.18	.022	.82	.032	.27
		2.................	.11	.023	.77	.061	.34
		3.................	.13	.021	.75	027	.39
237	M-7...	1.................	.17	.026	.78	.046	.37
		2.................	.C8	.031	.88	.048	.35
		3.................	.14	.024	.75	.034	.33
238	M-4...	1.................	.14	.032	.76	.048	.68
		2 and 3.................	.14	.032	.76	.048	.68
		2½.................	.02	.078	.17	.046	.39
239	M-5...	1 and 3.................	.16	.042	.67	.035	.37
		2.................	.18	.034	.69	.027	.45

1½-INCH STEEL CABLES

Serial No.	Manu-fac-turer	Position of wire in strand a	Si	P	C	S	Mn
248	M-2...	1, 2, and 3.................	0.10	0.034	0.77	0.043	0.49
249	M-9...	1 and 2.................	.19	.027	.86	.058	.36
		3.................	.24	.026	.96	.058	.48
		3.................	.20	.026	.80	.044	.25
250	M-1...	1, 2, and 3.................	.17	.028	.82	.063	.24
251	M-8 ..	1.................	.19	.032	.73	.039	.28
		2.................	.14	.029	.76	.032	.31
		3.................	.23	.026	.70	.048	.32
		3.................	.12	.026	.81	.055	.33
252	M-6...	1 and 2.................	.24	.033	.84	.038	.39
		3.................	.21	.043	.74	.040	.22
		3.................	.19	.048	.65	.028	.24
253	M-7...	1 and 2.................	.18	.035	.83	.044	.34
		3.................	.16	.023	.74	.032	.28
		3.................	.17	.028	.68	.065	.25
254	M-4...	1.................	.089	.027	.87	.057	.42
		2 and 3.................	.074	.036	.85	.043	.34
		2½.................	.013	.099	.15	.058	.36

a See Fig. 1 (d).

1. CHEMICAL ANALYSES OF STEEL

The results of the analyses are shown in Table 18. The steel for the chemical tests was taken from the wires of the 1¼-inch and 1½-inch cables. The wires selected for this purpose are designated in Fig. 1 (*d*) by the numerals 1, 2, 3, and 2½, 1 indicating the central wire, 2 the middle ring, 3 the outer ring, and 2½ the filler wires between the second and third rings. The serial numbers are those of Tables 11 and 12. The percentages of silicon, carbon, manganese, phosphorus, and sulphur as determined by standard methods are given in the table and illustrate the grades of steel used by the different manufacturers in meeting the provisions of the specifications. These results agree approximately with the results of similar tests made at the laboratory on plow-steel cables varying in diameters from five-eighths of an inch to 1¼ inches, but not included in the present series of tests. The wires are comparatively low in phosphorus and sulphur, except in the case of particular filler wires. The filler wires in some cases have less silicon and carbon, and the steel is softer and more ductile. The data from these tests may be compared with the tensile, torsion, and bending tests given later and which were made upon the same lengths of wires taken from these cables.

2. QUALITY OF FIBER IN ROPE CORES

The fiber of the rope cores of these 1¼-inch and 1½-inch plow-steel cables was submitted to microscopic examination. Three experts were also asked to pass judgment upon the quality of fiber used. (See Table 19.) There was complete agreement that the rope cores of cables submitted by manufacturers M–1 and M–2 were jute. One expert was of the opinion that the cores of the 1¼-inch cable submitted by M–6 and those of the 1¼-inch and 1½-inch cables submitted by M–4 were of manila, while the others were of the opinion that it was manila, with a mixture of sisal, which might, however, be "marketed" as manila. Two experts designated the cores of the 1¼-inch and 1½-inch cables of M–8 and the 1½-inch cable of M–5 as a mixture of istle and Mauritius hemp. The third was of the opinion that they contained, in addition, some New Zealand flax. All agreed that the core of the 1¼-inch rope of M–5 was composed of istle.

Manila fiber is generally considered to be the strongest. It has better resisting qualities against the action of sea water than the other fibers.[9] Jute is said to possess poor resisting qualities under the action of moisture. It is reported that sisal will not stand the action of sea water. While sisal is considered to possess the

---

[9] See Year Book, Department of Agriculture, 1903; ibid, 1909; also Mathew's Textile Fibres.

greatest tensile strength next to manila, it is not as flexible, and accordingly not as well adapted for hoisting or power transmission purposes. Istle has only comparatively recently been introduced for rope-core manufacture. Mauritius hemp is produced in large quantities on the Island of Mauritius and in Porto Rico and is relatively cheap in price. Both sisal and Mauritius hemp are comparatively unknown so far as the length of service for wire-rope purposes is concerned.

TABLE 19.—Quality of Fiber and Lubricant in Rope Cores

1¼-INCH ROPE CORES

Serial No.	Manu- facturer	Fiber in rope core, estimated	Lubricant and preservative

233...	M-2...	Jute...........................	Wood-tar product: Per cent
			    Tar.................................... 16.8
			    Tar acids............................. 39.9
			Oils:
			    Unsaponifiable....................... 41.7
			    Saponifiable......................... 1.6
			                    100.0

234...	M-3...	Manila.......................	Vegetable-tar product:
			    Tar.................................... 18.3
			    Tar acids............................. 34.8
			Oils:
			    Unsaponifiable....................... 41.7
			    Saponifiable......................... 5.2
			                    100.0

235...	M-1...	Jute	Wood-tar product:
			    Tar.................................... 9.8
			    Tar acids............................. 32.9
			Oils:
			    Unsaponifiable....................... 53.0
			    Saponifiable......................... 4.3
			                    100.0

236...	M-6...	Manila and small amount of sisal.	Probably a petroleum product consisting of vaseline, but darker and more opaque. It contains 2.2 per cent of saponifiable oils, calculated as lard oil, but containing no tar.
237...	M-7...	Manila and small amount of sisal.	Similar to above.
238...	M-4...	Manila (of poor quality)........	There was an insufficient amount for the obtaining of a sample for testing. It was apparently similar to that used on serial No. 254, 1½-inch diameter cable.
239...	M-5...	Istle.........................	Apparently a petroleum product consisting of vaseline, but more opaque and containing more vaseline. No tar was found. It had a fishy odor, which disappeared on removing the solvent (87° gasoline) used in extracting the material. Fat oil was found present to the extent of 0.60 per cent (calculated to glyceride of oleic acid).

TABLE 19.—Quality of Fiber and Lubricant in Rope Cores—Continued

1½-INCH ROPE CORES

Serial No.	Manufacturer	Fiber in rope core, estimated	Lubricant and preservative	
248...	M-2...	Jute.............................	Wood-tar product:	Per cent
			Tar....................................	17.8
			Tar acids................................	31.8
			Oils:	
			Unsaponifiable................................	41.9
			Saponifiable................................	8.5
				100.0
249...	M-9...	.....do.........................	Wood-tar product:	
			Tar....................................	20.5
			Tar acids................................	41.6
			Oils:	
			Unsaponifiable................................	32.9
			Saponifiable................................	4.8
				99.8
			Considerable rosin was present with the tar acids.	
250...	M-1...	.....do.........................	Wood-tar product:	
			Tar....................................	9.5
			Tar acids................................	32.9
			Oils:	
			Unsaponifiable................................	48.8
			Saponifiable................................	8.7
				99.9
251...	M-8...	Istle and Mauritius............	A petroleum product + a saponifiable oil. No tar was present.	
			Oils:	
			Unsaponifiable (petroleum oil)..................	87.1
			Saponifiable (calculated as glyceride of oleic acid)..	12.9
				100.0
252...	M-6...	Manila.........................	Wood-tar product:	
			Tar....................................	19.9
			Tar acids................................	19.3
			Oils:	
			Unsaponifiable................................	51.5
			Saponifiable................................	9.3
				100.0
			Graphite was found in the mixture, but was not determined, since it was mixed with dirt from handling.	
253...	M-7...	.....do.........................	Wood-tar product:	
			Tar....................................	25.4
			Tar acids................................	15.1
			Oils:	
			Unsaponifiable.	54.7
			Saponifiable................................	4.8
				100.0
			Graphite was present. (See note above.)	

TABLE 19.—Quality of Fiber and Lubricant in Rope Cores—Continued

1½-INCH ROPE CORES—Continued

Serial No.	Manu-facturer	Fiber in rope core, estimated	Lubricant and preservative	
254...	M-4...	Manila (of poor quality)........	Vegetable-tar product:	Per cent
			Tar...	40.8
			Tar acids....................................	7.5
			Oils:	
			Unsaponifiable.................................	50.7
			Saponifiable....................................	0.9
				99.9
255...	M-5...	Istle and Mauritius.............	A very dark heavy oil with strong fishy odor: Tar ..	5.0
			Oils:	
			Unsaponifiable (petroleum oil)..................	83.4
			Saponifiable (fish oil)..........................	11.6
				100.0

It is the opinion of the authors that manila fiber is the most suitable to use for rope cores for power transmission and hoisting cables on account of its tensile strength, its ability to stand the effects of moisture, and to better resist alternate bending stresses. It is more suitable for ships' rigging on account of its resisting qualities against sea water.

3. ANALYSES OF LUBRICANTS AND OF PRESERVATIVES FOR ROPE CORES

The materials here designated as lubricants and preservatives are not those usually added by the manufacturers of hoisting ropes, except in the case of the 1¼ and 1½ inch cables furnished by M-5 and M-8. (See Table 19.) These are used in manufacture and are found almost entirely on the rope cores. In some cases the amounts present were small, so that it was difficult to obtain a suitable sample for analysis.

The material reported as saponifiable in the wood and vegetable tar products may be either linseed oil, which is used by some manufacturers to thin the wood tars, or resinates, which are always present in the tars. At the time the analyses were made suitable apparatus was not available for separating the resin acids of the resinates and the mixed acid of the linseed oil. They have been calculated, therefore, as linseed oil. The error in doing so is slight.

4. TENSILE TESTS OF WIRES OF PLOW-STEEL CABLES

Numerous tensile tests of the wires were made. on the 10 000-pound testing machine shown in Fig. 13. Some of these tests will be discussed to draw attention to the typical physical characteristics of the wires. Table 20 shows the variations which may be expected in the tensile strengths and percentage elongations

TABLE 20.—Results of Tensile Tests of Wires of ⅝-Inch 6 by 19 Plow-Steel Cable, Serial No. 94

Strand No.	Position of wire [a]	Diameter of wire	Area of wire	Tensile strength		Elongation
		Inch	Inch^2	Pounds	Lbs./in.^{2}	Per ct.
1.......	1	0.0415	0.001353	327	241 700	3.3
	2	.0420	.001385	301	217 300	4.0
	3	.0410	.001320	344	260 600	4.0
	3	.0400	.001257	314	249 800	4.0
					[b] 242 400	3.8
2.......	1	.0450	.001590	378	237 700	3.3
	2	.0420	.001385	342	246 900	5.0
	3	.0410	.001320	335	253 800	5.0
	3	.0415	.001353	335	247 600	4.8
					[b] 246 500	4.5
3.......	1	.0450	.001590	380	239 000	4.8
	2	.0420	.001385	354	255 600	3.5
	3	.0420	.001385	343	247 700	5.0
	3	.0415	.001353	340	251 300	4.8
					[b] 248 400	4.5
4.......	1	.0450	.001590	364	228 900	4.5
	2	.0420	.001385	344	248 400	5.0
	3	.0410	.001320	314	237 900	5.3
	3	.0415	.001353	334	246 900	5.0
					[b] 240 500	5.0
5.......	1	.0455	.001626	358	220 200	5.0
	2	.0415	.001353	343	253 500	3.5
	3	.0410	.001320	346	262 100	3.3
	3	.0410	.001320	333	252 300	5.3
					[b] 247 000	4.3
6.......	1	.0450	.001590	360	226 400	5.0
	2	.0420	.001385	357	257 800	5.0
	3	.0410	.001320	337	255 300	5.0
	3	.0415	.001353	346	255 700	4.5
					[b] 248 800	4.9
					[c] 245 600	4.5

[a] See Fig. 1 (*d*).

[b] Weighted mean. Weights were taken proportional to the areas of wires.

[c] Average for strands = average for cable.

FIG. 13.—*Machine for obtaining tensile strengths of wires*

of the wires in a single strand of a cable. The strand was taken from a five-eighths inch plow-steel cable, serial No. 94 of Table 6. The position of the wire in a strand is indicated by numerals 1, 2, and 3. (See Fig. 1 (*d*).) The mean tensile strength of the wires was found to be 245 600 pounds per square inch, with a maximum departure from the mean of 28300 pounds per square inch. For the five-eighths-inch cables, serial Nos. 95 and 96, upon which similar tests were made of the wires, the mean values were found to be 253 100 and 252 300 pounds per square inch, respectively. The stresses for individual wires varied from 231 400 to 297 300 pounds per square inch in the first cable, and from 233 400 to 349 500 pounds per square inch in the second, the average departure from the means being much smaller. Table 21 gives in abstract the results of a large number of similar tensile tests on 1 ¼ and 1 ½ inch cables. It will be seen from the above data that the tensile strengths of the wires of the cables and strands are fairly uniform. An average variation from the mean of over 5 per cent is rather exceptional in the case of the wires of a single cable. The variations in the mean tensile strengths of different cables as determined on basis of the aggregate strength of the wires are usually larger than exists for the case of the wires of individual cables. This is explained probably by the fact that no two manufacturers submitted steel of precisely the same relative grade in meeting the stipulations of the specifications.

The elongations in percentage given in the tables were measured after rupture and give a good idea of the ductility of the wires used in the plow-steel cables. The average elongations are large when the high strength developed by the wires is considered. The elongations found for the wires of the five-eighths inch cable of Table 20 are unusual and rather too high to be representative.

The moduli of wires determined from tests upon wire taken from cables are not representative, because of the effect of initial curvature and twist in the wires from fabrication. In wire from reels previous tests have shown the modulus is about the same as for steel bars, 29 to 30 by 10^6. The moduli of the wires from the strands are much lower; values 22 to 26 by 10^6 pounds per square inch were found. For the same reason extensive measurements of elastic limits were not considered. In wire from the reels the elastic limit is high. The stress-strain curve for the cable of Fig. 11 is quite typical of the curves for wires.

TABLE 21.—Results of Tensile Tests of Wires in 6 by 19 Plow-Steel Cables

1/4-INCH CABLES

Serial No.[a]	Manufacturer	Position of wires in strand[b] (1=Central, 2=Middle ring, 3=Outer ring, 2½=Filler)	Diameter (Inch)	Tensile strength, 1000 pounds per square inch unit					Per cent of elongation					Per cent of reduction in area		
				Number of tests	Average	Maximum	Minimum	Average variation from mean per cent	Number of tests	Average	Maximum	Minimum	Average variation from mean per cent	Number of tests	Average	Average variation from mean per cent
233	M-2	1	0.091	3	228.7	263.0	199.0	10.0	2	2.88	3.00	2.75	4.2	3	35.1	10.2
		2 and 3	.081	18	244.1	274.0	192.0	7.0	14	2.64	3.50	2.00	12.1	18	43.4	5.8
234	M-3	1, 2, and 3	.084	18	239.4	262.0	221.0	2.9	12	2.42	3.13	1.25	16.1	18	43.9	5.0
235	M-1	1	.087	2	73.0	75.8	70.5	3.8	2	2.00	2.12	1.88	6.0	2	64.9	6.6
		2 and 3	.083	17	275.3	317.0	226.0	8.4	14	2.30	3.12	1.88	12.2	15	39.9	9.0
236	M-6	1	.115	2	252.0	253.0	250.0	....	2	1.76	1.88	1.63	....	2	45.5	....
		2	.058	18	305.0	325.0	286.0	2.6	12	0.78	0.88	0.62	14.1	17	42.3	5.0
		3	.101	17	263.0	273.0	255.0	1.5	16	2.40	3.12	1.87	10.5	17	43.8	2.7
237	M-7	1	.115	3	253.7	259.0	250.0	1.4	1	1.00	1.00	1.00	....	3	45.4	0.0
		2	.058	18	300.1	319.0	272.0	4.3	16	0.83	1.13	0.25	22.9	18	39.5	6.1
		3	.101	18	256.1	267.5	219.0	2.3	17	2.37	2.63	2.00	8.4	18	44.2	4.1
238	M-4	1	.091	1	226.0	226.0	226.0	....	1	2.00	2.00	2.00	....	1	41.2	....
		2 and 3	.083	18	240.0	253.0	230.0	2.5	14	2.23	2.88	1.50	10.8	18	35.0	14.0
		2½	.036	13	79.7	95.8	52.8	11.3	8	4.48	6.62	2.62	22.1	13	43.9	25.5
239	M-5	1 and 3	.103	18	225.7	239.0	215.0	2.2	9	2.27	2.50	2.13	5.7	18	44.5	5.8
		2	.054	18	212.0	231.5	191.0	4.3	13	2.09	2.50	1.75	7.7	18	53.1	6.0

1½-INCH CABLES

Serial	M-No.	Strand															
248	M-2	1, 2, and 3	0.098	18	253.5	281.0	201.0	8.0	15	2.07	3.00	1.50	15.5	18	45.9	5.0
249	M-9	1 and 2	.104	12	249.8	260.0	236.0	2.4	1	2.10	2.62	1.75	7.6	12	32.0	23.1
		3	.115	12	251.9	275.0	222.5	6.2	12	2.53	3.12	2.00	7.9	12	38.8	6.1
250	M-1	3	.078	12	245.1	259.0	215.0	3.4	12	2.07	2.38	1.75	8.7	12	42.5	4.5
		1, 2, and 3	.103	18	227.5	242.0	218.0	2.3	18	2.57	3.00	1.88	8.6	18	37.3	4.0
251	M-8	1	.110	3	220.5	223.5	217.0	1.0	2	2.38	2.38	2.38		3	37.0	4.6
		2	.102	1	232.4	252.0	219.0	3.2	6	2.29	2.62	1.88	8.7	11	38.1	6.8
		3	.115	12	216.4	235.8	173.0	4.7	1	2.62	3.00	2.25	3.4	12	34.7	7.8
		3	.080	1	240.7	283.0	223.0	5.8	10	2.05	2.88	1.75	8.8	1	38.6	6.0
252	M-6	1 and 2	.108	12	271.0	288.5	247.5	3.4	9	1.79	2.88	1.12	25.1	12	42.7	4.7
		3	.115	12	193.0	197.0	189.0	1.2	1	2.89	3.50	2.12	9.7	12	35.1	14.0
		3	.087	12	191.8	205.0	161.0	4.9	10	2.71	3.12	2.38	7.8	12	41.3	6.8
253	M-7	1 and 2	.108	12	271.3	293.5	255.0	2.6	12	1.95	2.38	1.38	9.7	12	42.9	5.7
		3	.115	12	192.1	201.0	186.0	1.8	10	3.01	3.25	2.75	4.0	12	35.9	14.8
		3	.087	12	190.0	201.0	164.0	4.9	10	2.79	3.12	2.50	6.8	12	45.0	6.2
254	M-4	1	.108	2	277.0	282.5	271.5	1.8						1	41.3	
		2 and 3	.100	18	238.0	253.5	204.0	2.9	16	2.33	3.12	1.50	15.5	18	42.9	5.6
		2½	.040	1	74.6	81.5	70.8	3.1	1	1.28a	1.48	1.11	6.3	1		
255	M-5	1 and 3	.124	18	a 210.6	224.5	201.0	2.4	14	2.95	3.38	2.62	6.1	18	47.0	4.0
		2	.062	18	232.4	240.0	221.5	1.8	8	2.14	2.50	1.88	7.0	18	47.1	7.4

a In serial Nos. 236, 237, 239, and 255 there were 9 wires in the middle ring (2) and 9 in the outer ring (3). The diameter of (2) is about half that of (3). This is known as the Seale construction. Only a few were included in the tests.

b See Fig. 1 (d). The filler wires (2½) are between the rings (2) and (3) in addition to the 19 wires of strand.

TABLE 22.—Results of Torsion and Bending Tests on Wires Taken from Plow-Steel Cables

¼-INCH CABLES

Serial No.	Manufacturer	Position of wires in strand [a] 1=Central 2=Middle ring 3=Outer ring 2½=Filler	Diameter	Torsion tests [b]									Bending tests [c]				
				Number of tests	Ultimate twisting moments, in inch pounds				Total number of twists of 360°				Number of tests	Initial tension units, 1000 pounds per square inch	Number of bends		
					Average	Maximum	Minimum	Average variation	Average	Maximum	Minimum	Average variation			Average	Maximum	Minimum
			Inch					Per cent				Per cent					
233	M-2	1	0.091										12	81.4	5.4	6.5	4.0
		2 and 3	.081	17	15.0	17.9	12.2	8.7	32.3	42.7	9.8	27.9	12	79.6	4.9	5.3	4.5
234	M-3	1, 2, and 3	.084	18	17.1	18.9	14.5	5.8	15.8	29.0	5.1	41.8	1	24.7	8.0	8.0	8.0
235	M-1	1	.087	18	15.1	17.5	13.3	7.3	8.4	17.8	2.5	34.5	11	91.3	3.7	4.5	3.0
		2 and 3	.083	2	54.3	55.5	53.0	2.4	25.3	25.6	25.0	1.2					
236	M-6	1	.115	18	6.4	7.1	5.9	5.2	17.7	29.5	4.5	18.1	11	101.7	4.7	5.5	4.0
		2	.058	17	33.3	35.0	31.5	2.7	13.2	26.2	3.6	31.1	11	87.7	6.2	6.8	6.0
		3	.101	1	57.0	57.0	57.0		20.1	20.1	20.1						
237	M-7	1	.115	18	6.3	7.1	5.6	4.8	14.6	25.5	7.5	15.8	12	100.0	5.1	6.0	4.5
		2	.058	18	28.9	32.1	23.3	5.2	12.5	20.9	5.4	26.4	11	85.4	6.2	7.0	5.0
		3	.101	1	21.0	21.0	21.0		14.0	14.0	14.0		1	75.3	5.5	5.5	5.5
238	M-4	1	.091	17	16.9	18.4	16.0	3.6	26.1	39.7	11.2	36.0	11	80.0	4.6	5.5	3.5
		2 and 3	.083	15	0.6	0.8	0.4	16.7	178.0	271.9	108.8	24.2					
		2½	.036	12	30.1	36.0	25.9	7.3	24.2	32.2	13.6	20.8					
239	M-5	1 and 3	.103	11	5.2	5.6	4.7	3.2	48.9	64.5	36.9	12.9	7	75.0	6.1	7.3	5.5
		2	.054										3	70.7	6.0	6.5	5.5

1½-INCH CABLES

Serial	Mark	Wires	Diam.	n									n				
248	M-2	1, 2, and 3	0.098	18	28.0	36.9	21.2	11.8	25.2	36.9	9.4	25.4	12	84.5	7.1	8.0	5.5
249	M-9	1 and 2	.104	12	34.0	36.0	32.0	3.5	18.6	31.2	5.3	55.4	10	83.3	5.3	6.0	4.5
		3	.115	12	45.4	48.5	40.5	4.2	13.9	28.9	2.2	69.8	10	84.0	5.0	6.3	3.0
		3	.078	1	12.8	14.6	11.1	7.0	34.0	41.2	13.8	21.5	9	81.7	4.9	5.5	4.0
250	M-1	1, 2, and 3	.103-	18	27.4	33.1	24.1	5.1	12.6	29.7	2.3	71.4	12	75.8	6.1	7.3	5.0
			.110														
251	M-8	1	.102	12	27.3	29.8	22.7	6.2	13.6	30.4	2.0	62.5	8	77.5	6.3	7.0	5.5
		3	.115	12	41.7	46.0	35.0	6.5	10.8	25.0	2.8	65.7	9	72.1	5.0	6.0	4.3
		3	.080	12	13.5	14.7	12.2	5.2	21.9	38.1	3.2	45.7	9	80.2	4.6	5.5	3.8
		1 and 2	.108	12	48.7	52.5	46.0	2.5	17.3	32.5	2.4	48.6	12	90.3	5.8	6.5	5.0
252	M-6	3	.115	12	40.5	47.0	36.5	3.7	16.0	26.6	5.0	42.5	10	64.3	5.3	6.5	3.0
		3	.087	12	15.3	17.5	10.4	8.5	32.8	35.4	28.0	6.1	9	63.9	4.7	5.5	4.0
		1 and 2	.108	12	48.4	51.5	43.0	3.7	15.9	28.5	6.4	28.3	10	90.4	5.7	6.0	5.0
253	M-7	3	.115	1	39.6	50.5	37.0	6.1	17.4	26.0	5.4	39.1	10	64.0	5.2	6.0	3.0
		3	.087	18	15.8	16.6	14.7	3.2	33.2	35.7	30.2	3.6	7	63.3	4.8	5.8	4.5
		1	.108														
254	M-4	2 and 3	.100	10	29.8	35.3	24.2	13.4	24.3	33.7	7.4	25.1	12	79.3	6.4	8.0	5.5
		2½	.040	18	0.9	1.0	0.8	3.3	165.6	216.7	117.7	17.8	8	24.9	13.0	15.3	10.5
255	M-5	1 and 3	.124		56.1	62.0	45.5	7.0	17.6	29.4	3.1	30.7					
		2	.062	18	7.2	8.2	6.2	6.9	39.2	54.4	7.6	26.3	8	77.5	7.0	8.0	6.0

a See Fig. 1 (d). The filler wires (2½) are between the rings (2) and (3) in addition to the 19 wires of the strand.

b An initial tension of 5000 pounds per square inch was applied except in the case of Serial Nos. 235 and 250.

c An initial tension was applied equal to one-third the tensile strength of the wire. The bends were through an arc of 90°. The load fell to 0.7 of initial tension after the first bend. The diameter of the jaws was equal to 4 diameters of the wire tested.

5. TORSION AND BENDING TESTS OF WIRES

Samples from the same lengths of wires of the 1¼ and 1½ inch diameter cables which were subject to tensile tests as given in Table 21 were also used in making the torsion and bending tests. The results of the tests are given in Table 22 for comparison.

The torsion tests were made in an Olsen wire-testing machine. The wire was clamped in the fixed head of the machine, the other end being permitted to pass entirely through the movable head and finally over a pulley. An initial tension of 5000 pounds per square inch was applied by means of a suitable weight. This initial load was placed on all wires, excepting the cables submitted by M–1. After applying the initial tension the wire was clamped in the movable head. The number of complete turns of 360° required to cause rupture was then read on a graduated arc on the head of the testing machine and is given in the table. The mean and extreme values of the accompanying twisting moment were measured. The force of torque was measured on a spring balance attached to the machine, the arm of the torque being 10 inches.

The bending tests were made on an Olsen wire-bending machine. One end of the wire to be tested was clamped to the weighing head. This is provided with a lever system by which the tension on the wire may be measured. The other end of the wire passes between two small steel blocks which are adjustable laterally for different diameters of the wires. The jaws of a set of these blocks are rounded to curves whose diameters are equal to about four diameters of the wires tested. The bending of the wire is applied by means of a swinging arc to which the wire is fastened by ordinary wedge grips after it passes through the rounded blocks of steel. The arc may be rotated alternately 45° about the small arcs of the blocks in each direction from the axis of the wire. An initial tension of approximately one-third the tensile strength was applied to each wire. After the first bend the load fell to about 0.7 of the initial value.

Tests of this nature are useful in showing any degree of uniformity the wires may possess with respect to their physical properties, but on account of the relatively large variations which are found it is difficult to draw conclusions as regards new wires. The results of the table show the comparative strengths of the different wires tested by this means. The positions of the wires in the cable are those indicated by the numerals on the section of the 6 by 19 construction shown in Fig. 1 (*d*).

VIII. LAW OF DISTRIBUTION OF STRESSES IN THE WIRES OF A CABLE

1. GENERAL ANALYSIS

If the wires possessed uniform elastic properties and were laid in parallel, as in the case of a large suspension bridge cable, the strength of the cable might be assumed to be equal to the aggregate strength of the component wires. It is evident from the more or less complex structure of ordinary cables as a result of the laying of the strands and wires that each wire does not resist stress equally. Some are, in fact, more stressed than others as the cable receives its load. It is common practice among engineers in forming an estimate of the strength of a cable, when direct tests are impracticable, to compute the aggregate strength of the wires and to multiply this by a coefficient. This factor in practice is variously estimated at from 80 to 95 per cent. It represents the ratio of the strength of the cable to the mean aggregate strength of its constituent wires. The ratio may be used to denote the efficiency of the construction. It depends on so many conditions—the construction and diameter of the cable, the tightness of laying of the wires, the degree of lubrication, the modes of end connections in tests, etc.—that a theoretical analysis is necessarily proximate.

The dynamics of stress distribution will be analyzed under certain ideal conditions.

(*a*) The surfaces of the wires are assumed frictionless—that is to say, infinitely smooth—so that any tangential forces acting along the surfaces of the wires as the result of their mutual contacts and slippings are zero.[10]

(*b*) The normal forces on the peripheries of a particular wire as a result of the mutual contacts and interactions from changes in curvature need not be considered. The mode of distribution has been defined for only the simpler cases, as where two or three wires are twisted together into a cord. These forces influence the degree of distortion of the wires. But by statics and the hypothesis (*a*) they present no components of force along the

---

[10] An individual wire, for example, may be conceived to trace a path between the others, as if they constituted a perfectly lubricated elastic-steel casing, the contact film also permitting freedom of elastic action of the wire under consideration. Let an element of length ds of this wire be taken. It may be considered as a small piston or a particle acted upon by forces in space; or, on the other hand, an element of the "casing" may be considered as a bead displaced by the strain through an infinitesimal arc of a tortuous curve. The particle is then subject to two conditions of restraint, $f_1 (x, y, z)$ and $f_2 (x, y, z)$, where x, y, z are the coordinates of position. The equilibrium of the particle in this case, both with and without friction, is discussed by Ziwet, Mechanics, Part II (1897), p. 137.

axis of the wires, except as occurs from changes of curvature in the wires caused by a couple acting along the strand. The theory of flexure shows that the stresses balance in pairs.

(c) When a strand or wire is considered as a type of spiral spring, and this is elongated by the load on the cable, the shearing stresses induced by torsion [11] and the flexure caused by twists and a change in curvature are neglected in comparison with the direct tension on the strand or wire from the loading. These stresses may be approximately estimated by the methods commonly in use for the combination and resolution of stresses, when considered in connection with the elongation and reduction data of the tables. It will be evident, however, that when a single wire or strand of the cable is considered *in situ* as such a spring, the elongation of the spring is only a few percentages at rupture and much less within the domain of elastic action. The force causing the elongation of the spring and the shears from the torque are small enough to neglect in comparison with the tension which is transmitted along the axis of the wire.

(d) The strands are assumed to carry equal portions of the load. A relative slipping of one strand, or a greater relative elongation in this strand from any cause, will throw more of the stress on others. The analysis thus implies perfect end connections, also uniform ductilities and other physical properties of the wires.

(e) The analysis is static. The loadings were applied at slow speed. There are doubtless slight relative slippings of wires at all stages of the loading, but the readjustments are taking place slowly. It is assumed in the general analyses of funicular problems

---

[11] Three neighboring points s_0, s_1, s_2 may be considered on the space curve traced by the central point of a strand or wire. The limiting position of the plane defined by these points when s_0 and s_2 are brought into coincidence with s_1 is called the osculating plane in the geometry of skew curves. For example, if a short length of the curled wire is laid upon a table the plane of the table approximately defines the position of the osculating plane with reference to the wire. The angle $\Delta\phi$ between two osculating planes taken, for example, at the points s_0 and s_2 represents the total torsion. The torsion $\frac{d\phi}{ds}$ represents the rate of turning of the osculating plane with reference to the increasing arc.

A plane may be taken perpendicular to the tangent of the curve at the point s_1 in question. All the normals to the curve are in this plane. The principal normal lies in the osculating plane, the binormal at right angles to it. The tangent, binormal, and principal normal are taken so as to form a right-handed system as coordinates of reference.

The curvature expresses the rate $\frac{d\psi}{ds}$ at which the tangent turns with reference to the increasing arc. The center of curvature lies on the principal normal at a distance from the point s_1 equal to the radius of curvature. The skew curve in the neighborhood of s_1 may be projected orthogonally on the osculating plane. The radius of curvature is equal to that of the plane curve thus defined at the point s_1 in question.

The shearing or angular straining is induced in the wires or the strand considered as a unit by a constant torque, viz, the load on the spiral times the radius of the spiral. There is in general a couple whose vector is axial. This causes the change in curvature and bending stress. The shearing and bending stresses here referred to are neglected in hypothesis (c) in comparison with the direct tensions on the wires and strands. (See E. Goursat, Cours d'Analyse (Hedrick's Trans.), 1, 1st ed., p. 468, 1904; A. E. H. Love, Theory of Elasticity, 2d ed., p. 396, 1906.)

that the structure is momentarily rigid when the conditions of equilibrium are applied. The planes of reference to be taken, in other words, are assumed to be conserved without relative movements of the wires with reference to the planes in the interval considered. Kinetic effects are not presented.

These considerations, even under the most favorable aspects, can not be fully realized in practice. There is never perfect lubrication nor uniformity of elastic properties and ductilities of the wires. There may be initial strains in the wires from twisting or tension in fabrication which are not present in the wires on the spools. A short specimen of cable with end sockets is not precisely equivalent mechanically to an indefinite length of cable, and the stress is not as uniformly distributed among the strands. An analysis according to the above conditions, nevertheless, defines an upper limit for the strength of the cable in terms of the strengths of the wires and expresses a certain desideratum of attainment in the making of standard tests under the most favorable conditions.

If a right section of a strand is considered at any point s_0 on its axis, the wires of any particular concentric ring of wires composing the strand intersect this plane at a common angle with the normal. Let the direction angle of a wire referred to the normal be θ_i $(i = 1, 2, 3 \ldots n)$, where the subscript i has different values according to the relative position of the ring with reference to the axis of the strand, the method of numbering having already been used for the special case of Fig. 1 (d). The areas of the wires and the tensile stresses for each ring, which by axial symmetry may be assumed uniform, will similarly be taken as a_i and p_i, respectively. The effective component of the combined stresses on a ring in the direction of the normal is $p_i \, a_i \cos \theta_i$. Considering the central wire as ring 1, the tension on the strand T_s is found by summing such expressions for all the rings or

$$T_s = \sum_1^n p_i a_i \cos \theta_i : (i = 1, 2, 3 \ldots n).$$

A neighboring section at a point s_1 may be taken at an element of distance Δs_0 along the axis of the strand. When the elastic stretch along the axis is $\delta(\Delta s_0)$ the corresponding stretch in the wires of the ith ring is given by the relation

$$\frac{p_i \Delta s_1}{E} = \frac{p_i \Delta s_0 \sec \theta_i}{E} = \delta(\Delta s_0) \cos (\theta_i + \epsilon_i).$$

E is the modulus of elasticity of the wires, Δs_i is the length of a wire of the ith ring intercepted by the neighboring planes of cross section, Δs_0 being taken very small, and ϵ_i is an angle depending upon the space coordinates of the skew curve generated by the wire. The angle ϵ_i expresses the angular decrement in the slope of the wire imposed as a result of the twist and change in curvature from the strain. It becomes infinitesimal for elastic strains, vanishing with Δs_0 for the condition of no load, and is neglected by hypothesis for a first and sufficient approximation.

If similar equations are written for each ring of wires, and the conditions for static equilibrium along the axis of the strand previously given are applied, there results the following group of equations:

(1) $$p_1 \sec^2\theta_1 = p_2 \sec^2\theta_2 = \ldots\ldots\ldots = p_n \sec^2\theta_n$$

(2) $$p_1 a_1 \cos\theta_1 + p_2 a_2 \cos\theta_2 + \ldots\ldots\ldots + p_n a_n \cos\theta_n = T_s$$

The solution of these equations gives for the first ring of wires, r_1:

(3) $$p_1 = \frac{T_s \cos^2\theta_1}{\left(\dfrac{a_1}{\sec^3\theta_1} + \dfrac{a_2}{\sec^3\theta_2} + \ldots\ldots\ldots + \dfrac{a_n}{\sec^3\theta_n}\right)}$$

and in general for the stress p_r in a particular ring r

(4) $$p_r = \frac{T_s \cos^2\theta_r}{\displaystyle\sum_{1}^{n} \frac{a_i}{\sec^3\theta_i}} : (i = 1, 2, 3\ldots\ldots\ldots n)$$

Theoretically the central or core wire of the strand from the above equation (3) receives the greatest tensile stress. Its angle of slope with the normal is zero and the square of the cosine is unity. The central wire, therefore, offers the most direct path for the transmission of stress in the strand. Its stress defines the upper limit of stress p in the strand, where $p = p_1 > p_2$, $p_3 \cdot \cdot p_r \cdot \cdot p_n$ by the law $p_r = p_1 \cos^2\theta_r$ as found from equations (3) and (4). The stress in the strand may, therefore, be represented in terms of the stress of its most stressed wire by the equation

(5) $$T_s = p \sum_{1}^{n} \frac{a_i}{\sec^3\theta_i} : (i = 1, 2, 3\ldots\ldots\ldots n)$$

If S represents the number of strands in the cable and ϕ is the direction angle between the tangent to the central line of the strand and the normal to a right section of the cable, the effective stress along the axis of the cable in terms of the stresses in the wires becomes

$$(6) \qquad T_o = S \, T_s \cos \phi = S \cos \phi \left(\sum_{1}^{n} \frac{a_i}{\sec^3 \theta_i} \right) p$$

where the coefficient of p, the stress in the most stressed wire, is to be determined for any particular type of construction.

2. ANALYSIS OF STRESS DISTRIBUTION IN 6 BY 19 PLOW-STEEL CABLES

Formula (6) will be applied to the case of plow-steel cables of 6 by 19 construction as an example of method. The treatment of other cases, as, for example, a 6 by 37 or 6 by 61 construction, is similar except that additional rings of wires are considered. In the 8 by 19 construction, 8 strands are considered instead of 6, and so on with the 6 by 7 and other types.

The aggregate cross-sectional areas of the wires in the different strands of standard concentric strand construction are related to each other as the terms of the sequence $-a$, 7a, 19a, 37a, 61a, $\ldots\ldots\ldots[3n(n-1)+1]a$, where n of the general term expresses the number of concentric rings forming the strand, and a is the mean area of a wire. The area of the cable in terms of the mean area of the wires is $S[3n(n-1)+1]a$, where S represents the number of strands. The respective total areas of the concentric rings are found by taking the first differences of the previous sequence.[12] The sequence for the aggregate areas of the wires of the different rings, exclusive of the first wire, is 6a, 12a, 18a,....6$(n-1)$ a. The constant difference of the suite is 6a, as in the case of the terms of an arithmetical progression.

Equation (6), taken in connection with the equation $p_r = p \cos^2 \theta_r$, may be used to calculate the stresses of the wires of each ring under a given load on the cable to a closer approximation. An opposite course will be pursued, however. Assuming that the safe or ultimate tension on the wires shall not exceed a certain amount as determined by tests of the wires, the effective component of the total stress along the axis of the cable will be computed for comparison with some of the results of the tensile tests of the cables.

---

[12] The areas and stresses of Tables 3 to 13 were computed upon the assumption of the wires being laid in parallel on account of practical considerations. The cross-sectional areas of the wires in the conventional sections of the cables of Fig. 1 are shown as assemblages of circles. More exactly in standard constructions these are in general ellipses of small eccentricities, the eccentricity varying with the coordinates of position of a wire, so that the practical analysis is not rigorous in a scientific sense.

Equation (6) becomes, in the case of the 6 by 19 construction, where only three terms of the sequence need be considered,

$$(7) \qquad T_o = 6 \cos \phi \, (\cos^3 \theta_1 + 6 \cos^3 \theta_2 + 12 \cos^3 \theta_3) \, pa$$

The value T_o as previously given is the load on the cable. The direction angle ϕ is the angle between the tangent to the generating helix of the tubular surface enveloping strand and the normal to the right section, and θ_1, θ_2, θ_3 are similar angles for the respective wires in the several rings taken with reference to the normals of the congruent sections of the strands. The term pa is the load limit on the most stressed or central wire of each strand. Numerical values of the terms are to be computed from the previous data.

The axes of these central wires intersect the right sections of the strands at radial distances $\dfrac{D}{3}$ from the axis of the cable, where D is the diameter of the cable. The mean lay or pitch of the helix from Fig. 2 may be taken as $7\frac{1}{2}D$. Therefore, the mean value of $\cos \phi$ becomes

$$\cos \phi = \cos \text{ arc tan } \frac{2\pi D}{3} \bigg/ \frac{15}{2} D = \cos \text{ arc tan } \frac{4\pi}{45} = 0.963$$

The wires of the second and third rings of the strands make a complete turn about the central axis of strand in a lay = approximately $3D$. The orthogonal projection on the transverse sections are accordingly $\dfrac{2\pi D}{15}$ and $\dfrac{4\pi D}{15}$, respectively, the space curves of the wires being treated as helices in the neighborhood of the point under consideration. Since θ_1 is zero, the numerical expression of the terms in the parentheses of (7) becomes

$$1 + 6 \cos^3 \left(\text{arc tan } \frac{2\pi}{45} \right) + 12 \cos^3 \left(\text{arc tan } \frac{4\pi}{45} \right)$$
$$= 1 + 5.88 \qquad\qquad + 10.72 = 17.60$$

Accordingly, there results for T_o

$$T_o = 6 \times 0.963 \times 17.60 \, pa = 101.7 \, pa$$

Since the aggregate cross-sectional areas of the wires times the allowable stress is $114 \, pa$, the ratio of the strength of the 6 by 19 cable to the total available strength of its wires, as deduced from theoretical considerations, becomes 0.892. This ratio will be called, for brevity, the efficiency of the construction.

3. CALCULATION OF EFFICIENCIES FROM DATA OF TESTS

The actual efficiencies of plow-steel cables of five-eighths inch, 1¼ inch, and 1½ inch diameters have been calculated from the data obtained from tests of about 600 wires from these cables. An abstract of the results on the wire tests for the 1¼ and 1½ inch specimens is given in Table 21. The data for the five-eighths inch specimens were obtained as illustrated in Table 20. The mean area of each size of wire entering the construction of the specimens was calculated. The average breaking stresses observed for different wires recorded in the tables were multiplied by the respective areas and added to give the mean aggregate strength of the wires in the cable. The observed maximum loads on the cables as given in Tables 6, 11, and 12 were divided by the aggregate strengths of the wires to obtain the efficiencies recorded in Table 23.

The average efficiency of the cables as calculated from the above tests is 81.3 per cent, or 7.9 per cent lower than that found from the theoretical considerations. The authors are of the opinion that this discrepancy is to be explained mainly by the fact that the wires are not all of the same ductility and do not possess precisely the maximum strengths. Accordingly some wires are stretching more than others near the maximum load. The wires with less percentages of elongation tend to break first. The result is that the strands near the breaking point are eventually unequally stressed. Tables 3 to 12 show that the average number of strands breaking is from two to three, this indicating that the strands are not stressed and distorted uniformly at rupture. The maximum stresses of the weaker strands thus determine the maximum load. If all strands could be arranged to break simultaneously probably the strength of the cable would be increased somewhat.

It was shown previously that the stress p_r for any particular ring of wires is given in terms of the stress for the central or most stressed wire by the formula $p_r = p \cos^2 \theta_r$ where θ_r is the inclination of the wires to the normal to the right section of strand. Tables 21 and 22 show that the elongations are not graded proportionally to the stress in the ring. It seems reasonable, therefore, to believe that if it were possible to do this, it would tend to equalize the stresses on the strands to some extent. For example, the load on the central wire is the greatest since it has zero slope to the normal to the right section of strand and, therefore, offers the most direct path for the transmission of stress. Other things being equal, it tends to break first. If the percentages of elongation in this wire were slightly greater than in the surrounding wires, it would stretch proportionately a greater amount before breaking and thus the other wires would tend to equalize the stress distribution over the cross section of the strand, and similarly with the remaining wires. It was previously stated that certain interior wires of the cable usually break first with sharp metallic rings. This shows some verification of the theory presented.

TABLE 23.—Calculated Efficiencies of 6 by 19 Plow-Steel Cables

⅝-INCH CABLES

(a) Serial No. of tables	(c) Observed maximum load on cable	(d) Observed mean maximum stresses in wires	(e) Aggregate sectional areas of wires	(f) Efficiency in per cent equals ratio of strength of cable to mean strength of wires × 100 $E = \dfrac{(c)}{(d) \times (e)} \times 100$
	Pounds	Lbs./in.²	Inch ²	
94	30 200	245 600	0.1592	77.2
0 a	30 200	270 000	.1307	85.7
0 a	29 700	272 500	.1307	83.4
95	31 260	253 100	.1485	83.1
96	29 550	252 300	.1485	78.7
Average....................				81.6

1¼-INCH CABLES

233	108 000	237 000	0.6020	75.7
234	123 800	239 400	.6270	82.5
235	119 000	266 000	.6180	72.4
236	130 000	272 000	.6350	75.4
238	126 700	234 500	.6590	82.0
239	125 500	224 000	.6220	90.2
Average				79.7

1½-INCH CABLES

248	180 900	253 500	0.8600	82.9
249	176 500	249 000	.9030	78.3
250	188 300	227 500	.9500	87.3
251	148 000	227 300	.9060	72.1
252	183 000	226 000	.9730	83.3
253	184 500	223 000	.9730	85.1
254	190 800	232 000	.9480	86.7
255	163 500	214 600	.8880	85.7
Average				82.7
Grand average...................				81.3

a Not listed.

Under the conditions of working stresses which are commonly taken when there is little impact or bending at about one-fifth the maximum stresses developed by tests, the effects of unequal elongations in the wires at rupture are absent. It may be assumed from the relative constancy of the modulus of elasticity of the uncoiled wires that the distribution of stresses in the wires is then more nearly in accord with the theoretical calculation which has been given when the cable is well lubricated. Since the strands will behave as a set of coaxial helical springs of equal pitch interwinding one within another, the elastic stretch will never be precisely uniform. The wires do not bed on each other in the different strands in quite the same way, as is shown indirectly from the evidence presented by the tables of elongations and reductions. As the load is removed the recovery is not uniform for each strand, and, similarly, when the wires themselves are considered as springs. The mutual interactions at the peripheries will be present. If, however, the cable is frequently lubricated, the frictions brought into play by the reaction of one wire or strand on another will be kept to a minimum, thus premitting the strands and wires to glide over each more freely. Each strand will tend to carry its proper proportion of the load where there is more opportunity of readjustment between the components. The cable will be better fitted to resist alternate bending stresses which cause broken wires when it is properly lubricated.

4. IMPORTANCE OF LUBRICATION OF HOISTING CABLES

The authors have found by examination and tests of a number of used cables that there is often considerable impairment and lowering of the life of wire rope through insufficient lubrication. Many wires are found to be corroded. The coefficient of friction is then materially increased. The opportunity for readjustment and sliding when there is a tendency to overstressing is not present. The rope core deteriorates, becoming dry and inelastic and easily disintegrable. The wires are thus broken more easily, even when the sheave is of appropriate curvature, since the cable acts like a beam rather than a flexible cord. The corroding pits the wires and reduces their available sectional areas. Improper stress distributions tend to dent the wires and lower their strength in much the same way as the corroding action. In cables from mine hoists the lubricant is sometimes found to have congealed through lack of proper attention into a hardened matrix. As the stiffness and modulus of the cable is then materially increased, the bending stresses are larger than under normal conditions. Many broken wires are found in such cases. Systematic lubrication of hoisting cables at stated intervals will prolong their life in service and insure more favorable stress distributions among the wires.

IX. SUMMARY AND CONCLUSIONS

The paper presents the results of tests of 275 wire ropes submitted by American manufacturers to fulfill the specifications of the Isthmian Canal Commission. The selections of the test specimens were made by Government inspectors. The ropes were of diameters ranging from one-fourth of an inch to $1\frac{1}{2}$ inches, a few being of larger diameters up to $3\frac{1}{4}$ inches. Over half the specimens were plow and crucible-cast steel hoisting rope of 6 and 8 strands of 19 wires each. The remainder were guy and tiller ropes of 6 strands of 7 wires and 6 strands of 42 wires each.

1. RECAPITULATION OF STRUCTURAL DATA

The linear dimensions of the wires, strands, and rope cores were found from measurements to be proportional to the diameters of the cables. The diameters of strands and rope cores were generally one-third the diameters of the cable, the cores of the 8 by 19 plow-steel rope being slightly larger. The mean pitch or lay of a strand was approximately $7\frac{1}{2}$ times the diameter of the cable. The mean lay of the wires was approximately $2\frac{3}{4}$ times the diameter of the cable. The mean diameters of the wires are given approximately by the equation

$$d = \text{diameter of wires}; \quad D = \text{diameter of cable}$$
$$N = \text{number of wires in outer ring of strand}$$
$$d = K\frac{D}{N+3}: \quad K = \begin{cases} 1.0 \text{ for hoisting and guy rope} \\ .8 \text{ for flexible hoisting rope} \\ .33 \text{ for tiller rope.} \end{cases}$$

The mean aggregate sectional area of the wires in a cable in terms of its diameter is given approximately by the formula

$$A = C D^2: \quad C = \begin{cases} 0.41 \text{ for 6 by 19 plow-steel rope} \\ .38 \text{ for 6 by 19 crucible-cast steel rope} \\ .38 \text{ for 6 by 7 guy rope} \\ .35 \text{ for 8 by 19 plow-steel rope} \\ .26 \text{ for 6 by 42 iron tiller rope.} \end{cases}$$

The sectional area of the steel in a cable in terms of the mean area of its wires is given by the formula

$$A = S[3n(n-1)+1]a: \quad \begin{cases} A = \text{area of steel} \\ S = \text{number of strands} \\ n = \text{number of concentric rings} \\ a = \text{mean area of a wire.} \end{cases}$$

2. RECAPITULATION OF THE RESULTS OF TESTS OF CABLES

It was found when the observed maximum loads were platted as functions of the diameters of the cables of each class that the lower boundary of the field comprising these observations could be expressed within fairly close limits by the formula

$$\text{Load} = C \; 75 \; 000 \; D^2$$

$$D = \text{diameter of cable}$$

$$C \begin{cases} = 0.9 \text{ to } 1.1 \text{ ; mean about } 1.0 \text{ , } - \text{Plow steel 6 by 19 cables} \\ = .8 \text{ to } 1.00; \text{ mean about } .85, - \begin{cases} \text{Plow steel 8 by 19} \\ \text{Crucible-cast steel 6 by 19} \end{cases} \\ = .3 \text{ to } .45; \text{ mean about } .35, - \begin{cases} \text{Tiller rope 6 by 42} \\ \text{Guy rope 6 by 7} \end{cases} \end{cases}$$

The specifications and the standard strengths of the manufacturers' were found in general to agree quite closely with the loads defined by these lower boundaries. The arithmetical means of the observed maximum loads from the tests were usually about 5 to 12 per cent higher than the minimum values recorded, depending upon the particular grade of steel used by the manufacturer in meeting the requirements of the specifications.

The mean values of the observed maximum stresses found for the different classes of cables when platted in curves showed a general correspondence with similar curves platted from stresses figured from the maximum loads given by the specifications and the standard strengths of the manufacturers. The smooth curves following the general trend of the platted observations were of the type found in tests of wires of different diameters. The relatively high maximum stresses found for small cables were attributed to the greater strengths of the wires as a result of wire drawing.

The observed unit elongations under cumulative stresses showed some irregularities for different cables of the same class and diameter. The elongations were nearly proportional to the stresses in a particular cable. The calculated moduli of the cables varied from 3 by 10^6 to 9 by 10^6 pounds per square inch of cable section. While this is increased by service, it is believed the limit of 12 000 000 pounds per square inch used in the calculation of bending stresses is ample.

3. RECAPITULATION OF RESULTS OF TESTS FOR QUALITY OF MATERIAL

The tensile strengths of wires in a strand were quite uniform. The maximum elongations were relatively high, considering the strength of the wires. The strengths of wires in cables of the same class and diameter were quite uniform when the steel was of the same grade. The relatively large departures from the mean in a group of cables of the same class may be attributed to the fact that the chemical constituents are not identical in all cases, but depend upon the particular type of steel submitted by the manufacturers in meeting the provisions of the specifications. The variations in the strengths of the cables were larger than would have occurred if all specimens of a class had been selected from the same manufacturer and had they been constructed of the same grade of steel.

The fiber used for rope cores in the hoisting cable of $1\frac{1}{4}$ and $1\frac{1}{2}$ inches diameter was estimated to be manila, jute, or istle. In some cases mauritius had been mixed with the istle. Manila fiber is considered to be most efficient for power transmission and hoisting cables, on account of its greater strength and ability to resist alternate bending. It is probably more efficient for ship rigging and like purposes, where the cables are exposed to sea water.

The preservative and lubricants used were wood or vegetable tar combined with petroleum oil or other petroleum products. Experience from tests of cable used in practice shows that the cables when bent over sheaves should be lubricated at frequent intervals to prevent corrosion of the wires and rigidity and drying out of the core, and to insure more freedom of action of the strands and wires in adjusting themselves in resisting direct and bending stresses. Otherwise the strength is impaired by corrosion and denting of the wires from irregular stress distributions.

The efficiency of a cable in developing the strength of the wires depends theoretically upon the construction and lays of the strands and wires. The effective component along the cable axis was found to be about 89 per cent of the working stress in the wires for 6 by 19 plow-steel cables. The mean component developed at rupture was found to be about 8 per cent less than the theoretical value computed on the basis of an elastic behavior of the material. This was attributed to different degrees of ductility possessed by the wires and some variation in their strengths. It is the opinion that the efficiency of different constructions will be in closer agreement with the computed efficiency when working stresses are not exceeded.

WASHINGTON, June 29, 1918.

CPSIA information can be obtained
at www.ICGtesting.com
Printed in the USA
BVHW08*1009031018
529155BV00011B/414/P